SpringerBriefs in Applied Sciences and Technology

SpringerBriefs present concise summaries of cutting-edge research and practical applications across a wide spectrum of fields. Featuring compact volumes of 50 to 125 pages, the series covers a range of content from professional to academic.

Typical publications can be:

- A timely report of state-of-the art methods
- An introduction to or a manual for the application of mathematical or computer techniques
- A bridge between new research results, as published in journal articles
- A snapshot of a hot or emerging topic
- An in-depth case study
- A presentation of core concepts that students must understand in order to make independent contributions

SpringerBriefs are characterized by fast, global electronic dissemination, standard publishing contracts, standardized manuscript preparation and formatting guidelines, and expedited production schedules.

On the one hand, **SpringerBriefs in Applied Sciences and Technology** are devoted to the publication of fundamentals and applications within the different classical engineering disciplines as well as in interdisciplinary fields that recently emerged between these areas. On the other hand, as the boundary separating fundamental research and applied technology is more and more dissolving, this series is particularly open to trans-disciplinary topics between fundamental science and engineering.

Indexed by EI-Compendex, SCOPUS and Springerlink.

Abdallah Hamed

Speckle Imaging Using Aperture Modulation

 Springer

Abdallah Hamed
Department of Physics
Faculty of Science
Ain Shams University
Cairo, Egypt

ISSN 2191-530X   ISSN 2191-5318 (electronic)
SpringerBriefs in Applied Sciences and Technology
ISBN 978-3-031-58299-8   ISBN 978-3-031-58300-1 (eBook)
https://doi.org/10.1007/978-3-031-58300-1

This Springer imprint is published by the registered company Springer Nature Switzerland AG
The registered company address is: Gewerbestrasse 11, 6330 Cham, Switzerland

Paper in this product is recyclable.

Dedications
To my parents and my wife

Preface

Even since the invention of lasers in 1960, there has been a renaissance in the field of optics and the field of optical electronics. Optical and optical electronics are now being applied in all branches of science and engineering. The famous book *Introduction to Fourier Optics and Holography* by Goodman (1968) and *Laser Speckle and Applications in Optics* by Francon (1978), followed by my recent publications on the subject led to the presentation of the book *Modulated Apertures and Resolution in Microscopy* by Hamed AM (2023) followed by the present book on *Speckle Imaging Using Aperture Modulation*. The content of the book extracted from my recent publications adds information on speckle imaging due to the modification that occurred in pupil distribution. This book is intended for graduate students in optical sciences.

The object of drafting a book on speckle imaging using pupils with different distributions is outlined below. When the circular aperture of uniform transmittance was replaced by modulated cracks the point spread function (PSF) changed, and the cutoff spatial frequency was changed leading to resolution improvement in the formed image using a fixed diffuser.

The book is composed of nine chapters about the formation of speckle images using modulated apertures and laser spectral Voigt distributions. In these chapters, we calculated the impulse response or the PSF corresponding to the apertures and hence obtained the resolution in the formed speckle images. The recognition of the direction of new apertures from elongated speckle images is outlined in Chap. 1. Speckle images using Gaussian and graded index apertures are investigated in Chap. 2. In addition, the contrast of the formed speckle images is outlined in Chaps. 3, 4, and 6. Among the investigated modulated apertures there is a Hamming linear distribution as described in Chap. 5 and others have concentric black and white hexagonal pupils as described in Chap. 7. The speckle images with irregular apertures are discussed in Chap. 8. Finally, in Chap. 9 we computed the intensity distribution for the new Hermit Gaussian annular aperture and plotted it for different transverse modes. Second, we calculated the speckle images corresponding to these apertures using the FFT technique which shows the dependence of the speckle pattern upon the beam nonuniformity.

Cairo, Egypt   Abdallah Hamed

Contents

Chapter 1
Recognition of the Direction of New Apertures from Elongated Speckle Images

1.1 Introduction

Long elongated speckle images were obtained using mechanical scanning of the static speckle pattern [1]. The author presented a technique of spatially oriented speckle-screen encoding to improve the grating encoding technique for white-light image processing. Additionally, an artificial screen composed of small strips was photographed on a high-resolution film designed to obtain elongation ten times the average grain size of natural speckles [2]. In a recent publication by the author [3], numerical elliptical apertures of small elliptic shapes were analyzed and the Fourier transform was used to obtain speckle images of diffusers modulated by these elliptic apertures.

An approach for determining the roughness of engineering surfaces is the result of the speckle elongation effect. The laser speckle pattern, arising from light scattered from rough surfaces that are illuminated by polychromatic laser light, is detected in the far-field region. The incoherent superposition of these light intensities and angular dispersion cause speckle elongation [4]. This is characterized by increasing speckle widths and leads to a radial structure of the speckle patterns. With increasing surface roughness, the elongation is increasingly replaced by the decorrelation of the monochromatic speckle patterns for the different wavelengths. Such effects are detected with the CCD technique and analyzed by local autocorrelation functions of intensity fluctuations that are calculated for different areas of the speckle patterns. Hence, the autocorrelation method is applied to process laser speckle patterns. The relationships between the surface roughness and speckle elongation and between the correlation length of the autocorrelation function can be obtained. Consequently, a high surface roughness can be measured [5]. An oriented photographic diffuser is used to record an elongated speckle pattern. It was found that the contrast transfer when gratings were imaged through the slits in the diffuser was greater than that when imaging through a circular pinhole of comparable dimensions [6]. An auto-correlation algorithm for speckle size evaluation has been investigated [7–10]. The

© The Author(s), under exclusive license to Springer Nature Switzerland AG 2024
A. Hamed, *Speckle Imaging Using Aperture Modulation*,
SpringerBriefs in Applied Sciences and Technology,
https://doi.org/10.1007/978-3-031-58300-1_1

authors measured the average speckle size from the autocovariance function of the digitized intensity speckle pattern. Spatial characteristics such as speckle size can be used to measure the roughness of surfaces [11, 12]. An important remark taken into consideration during the recording of speckle data is that the speckle size must be greater than the pixel size of the CCD camera to resolve variations in speckle intensity [13, 14].

In this chapter, two sharp elliptical apertures with different orientations were proposed [15], and corresponding elongated speckle images were obtained. Additionally, three different models are numerically fabricated, and the corresponding speckle images are computed and plotted. Finally, the reconstructed models of ellipses and the autocorrelation of these models are plotted. The sharp ellipses considered in this study have a semimajor axis ten times greater than the semiminor axis considered in recent work [3] in which the semimajor axis is equal to the semiminor axis.

1.2   Theoretical Analysis for Speckle Imaging

1.2.1   Formation of Speckle Images Using Diffusers Modulated by Different Sharp Elliptic Apertures

In this chapter, a simple ellipse of the semimajor axis ten times the semiminor axis is investigated. We assume that this sharp ellipse is represented as follows:

$$f(x, y) = 1; \quad \frac{x^2}{a^2} + \frac{y^2}{b^2} \geq 1 \ \text{ with } \ b = 0.1\,a \tag{1.1}$$

The numerical ellipse is made from pixels with dimensions 1024×1024 and is represented as follows:

$$f(m\,\Delta x, n\,\Delta y) = 1, \quad (M, N) = (1024, 1024) \tag{1.2}$$

Δx and Δy are unity.

A diffuser of random distribution has the same dimensions as the above-described elliptical aperture constructed as:

$$d(m\,\Delta x, n\,\Delta y) = \text{rand}(m\,\Delta x, n\,\Delta y) \tag{1.3}$$

Consider a coherent imaging system with laser uniform illumination incident upon the object (the diffuser) followed by the manufactured sharp elliptic aperture. In this case, we can write the complex amplitude transmitted as follows:

$$c(m\,\Delta x, n\,\Delta y) = f(m\,\Delta x, n\,\Delta y) \cdot d(m\,\Delta x, n\,\Delta y) \tag{1.4}$$

Hence, the diffuser followed by the aperture and then the Fourier transforming lens give the complex amplitude in its front focal plane as follows:

$$\tilde{c} = \tilde{f}\left(\frac{m}{\Delta x}, \frac{n}{\Delta y}\right) * \tilde{d}\left(\frac{m}{\Delta x}, \frac{n}{\Delta y}\right) \tag{1.5}$$

$\tilde{f} = \text{F.T.}[f(m\Delta x, n\Delta y)]$, $\tilde{d} = \text{F.T.}[d(m\Delta x, n\Delta y)]$, and $\tilde{c} = \text{F.T.}[c(m\Delta x, n\Delta y)]$.

The symbol $(*)$ represents the convolution product.

Equation (1.5) indicates that the modulated elongated speckles formed from the convolution product of both the Fourier spectrum of the diffuser and the sharp elongated aperture. The direction of the speckle elongates normally to the semimajor axis of the elliptic aperture.

1.2.2 Effect of Aperture Tilting on Elongated Speckles

Assume that the aperture is misaligned making an angle α with the normal to the aperture plane (x, y). In this case, the sharp elliptic aperture is written as follows:

$$f_{\text{tilted}} = \exp[jky\sin(\alpha)]f(x, y) \tag{1.6}$$

Hence, the coherent light emitted from the laser beam is incident upon the tilted aperture described by Eq. (1.6) and the transmitted light is incident upon the diffuser described in the preceding Sect. 1.3 giving this complex amplitude:

$$c_{\text{tilted}}(m\Delta x, n\Delta y) = f_{\text{tilted}}(m\Delta x, n\Delta y) \cdot d(m\Delta x, n\Delta y) \tag{1.7}$$

Substituting Eq. (1.6) into Eq. (1.7), we rewrite Eq. (1.7) as follows:

$$c_{\text{tilted}}(m\Delta x, n\Delta y) = \exp[jky\sin(\alpha)] \cdot f(x, y) \cdot d(m\Delta x, n\Delta y) \tag{1.8}$$

Applying the Fourier transform operation and making use of the convolution operation, we obtain:

$$\tilde{c}(\text{tilt.apert.}) = \tilde{f}\left(\frac{m}{\Delta x}, \frac{n}{\Delta y}\right) * \tilde{d}\left(\frac{m}{\Delta x}, \frac{n}{\Delta y}\right) * \delta\left(\frac{m}{\Delta x}, \frac{n - f\sin\alpha}{\Delta y}\right) \tag{1.9}$$

δ is the Dirac delta distribution.

This delta function displaces the whole speckle pattern by an amount equal to f $\sin\alpha$ in the direction conjugated to the y-direction since f is the focal length of the Fourier transform lens. Hence, for a tilted aperture, the complex amplitude in the Fourier plane is written as follows:

$$\tilde{c}(\text{tilt.apert.}) = \tilde{c}\left(\frac{m}{\Delta x}, \frac{n - f \sin \alpha}{\Delta y}\right) \tag{1.10}$$

1.2.3 Autocorrelation Algorithm for Speckle Size Evaluation

The average speckle size of a speckle image is estimated by calculating the autocorrelation function of the digitized intensity speckle pattern as follows.

If $I(x_1, y_1)$ and $I(x_2, y_2)$ represent the intensities of two points in the imaging plane (x, y), the intensity autocorrelation function is defined by the following equation:

$$C_I(\delta x, \delta y) = \langle I(x_1, y_1) I(x_2, y_2) \rangle \tag{1.11}$$

where $\delta x = x_1 - x_2$, $\delta y = y_1 - y_2$, and <> corresponds to a spatial average.

The autocorrelation function corresponds to the normalized autocorrelation function of the intensity which has a zero base and its full width at half maximum (FWHM) supplies a reasonable measurement of the average width of a speckle [6]. Using the autocorrelation function method to calculate the average speckle size requires sufficient sampling of speckles in an image for a reasonable statistical evaluation.

MATLAB was used [8] to compute the autocovariance of the speckle image. The calculated autocorrelation functions are shown, and the FWHM of the calculated function gives the average speckle size of the speckle pattern.

1.2.4 The Reconstruction Process and the Autocorrelation of Elliptic Apertures

The complex amplitude of modulated speckle images using the above apertures is given by Eq. (1.5) as follows:

$$\tilde{c}(u, v) = \tilde{f}\left(\frac{m}{\Delta x}, \frac{n}{\Delta y}\right) * \tilde{d}\left(\frac{m}{\Delta x}, \frac{n}{\Delta y}\right) = \tilde{f}(u, v) * \tilde{d}(u, v) \tag{1.12}$$

$(u, v) = \left(\frac{m}{\Delta x}, \frac{n}{\Delta y}\right)$ are the reduced coordinates in the speckle plane (Fourier plane).

The reconstruction of the different apertures is obtained by running the inverse Fourier transform on Eq. (1.5) to obtain:

$$R(x', y') = \text{F.T.}^{-1}\left\{\tilde{f}(u, v) * \tilde{d}(u, v)\right\} \tag{1.13}$$

(x', y') are the Cartesian coordinates in the imaging reconstruction plane.

Making use of the properties of the Fourier transform and convolution, we obtain the following result:

$$R(x', y') = f(x', y') \cdot d(x', y')$$ (1.14)

Hence, in the imaging plane, we localize the aperture image affected by noise originating from the diffuser function.

The autocorrelation function of the different apertures obtained by running the Fourier transform on the intensity of the speckle image is as follows.

First, we obtain the speckle intensity as the modulus square of the complex amplitude of the speckle Eq. (1.12) as follows:

$$I(u, v) = \left|\tilde{c}(u, v)\right|^2 = \left|\tilde{f}(u, v) * \tilde{d}(u, v)\right|^2$$ (1.15)

By operating $F.T^{-1}$ over Eq. (1.15), we obtain the autocorrelation function of the multiplication product of $f(x', y').d(x', y')$ as follows:

$$c(x', y') = F.T.^{-1}\{I(u, v)\} = F.T.^{-1}\left\{\left|\tilde{f}(u, v) * \tilde{d}(u, v)\right|^2\right\}$$
$$= \left[f(x', y') \cdot d(x', y')\right] * \left[f^*(x', y') \cdot d^*(x', y')\right]$$ (1.16)

1.3 Results and Discussion

The diffuser is multiplied by the sharp elliptical aperture of the semimajor axis by ten times the semiminor axis. A matrix of dimensions 1024×1024 pixels is considered for this diffuse aperture. The elongated speckle pattern was obtained by running the FFT algorithm on the diffuse aperture. The elongation of the speckles is normal to the major axis of the ellipse shown along the x-direction. Hence, the elongation is directed along the y-direction. Figure 1.1 shows the diffuse sharp ellipse with its major axis, for which $\theta = 45°$ with respect to the x-axis and the corresponding elongated speckle with elongation orthogonal to the pupil major axis. The figure shows that the direction of the ellipse is recognized from the direction of the elongated speckle where both must be orthogonal to each other.

All the elongated speckle images shown in the above Figs. 1.1, 1.2, 1.3, 1.4, 1.5 have dimensions of 256×256 pixels.

A model of a pupil in the form of an airplane superimposed over a diffuser with dimensions of 2048×2048 pixels was fabricated as shown in Fig. 1.2a. The corresponding speckle image obtained by running the FFT showed a specific elongation along the (x, y) plane, as shown in Fig. 1.2b.

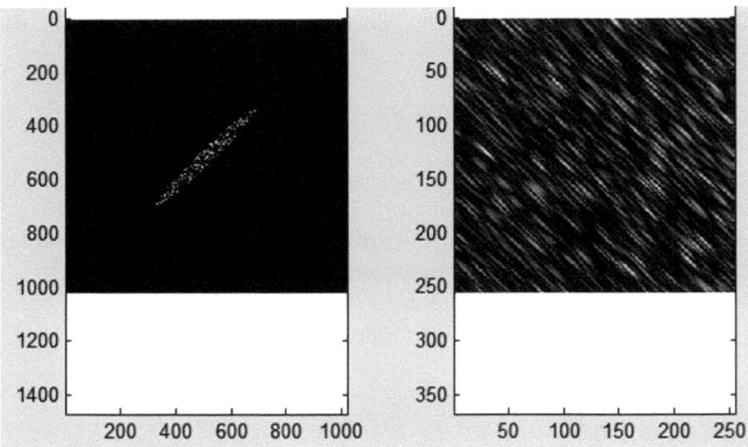

Fig. 1.1 On the left, an elliptical aperture with an angle of 45° and x-axis and matrix dimensions of 1024 × 1024 pixels is shown. On the right is the corresponding speckle image elongated normally to the semimajor axis with matrix dimensions of 256 × 256 pixels. The elliptical aperture is superimposed over the diffuser

The profile shape of the autocorrelation intensity is shown in Fig. 1.3a, along the x-axis corresponding to the elongated speckles in Fig. 1.2b for an airplane pupil of 2048 × 2048 pixels. Additionally, the autocorrelation profile along the y-axis is shown in Fig. 1.3b. The speckle size is computed by taking the full width at half maximum along the x-direction, referring to Fig. 1.3a as follows: FWHM = Δx = 5 pixels and along the y-direction, referring to Fig. 1.3b as Δy = 8 pixels. Hence, the average speckle size in the x-direction is calculated as follows:

σ_x = (4.5 mm/512 pixels) (5 pixels) = 44 μm.

The average speckle size in the y-direction is computed as follows:

σ_y = (3.6 mm/512 pixels) (8 pixels) = 56 μm.

For comparison, the average speckle size for a circular aperture with a radius of 128 pixels is obtained as follows:

σ_x = (4.5 mm/1024 pixels) (10 pixels) = 44 μm, σ_y = (3.6 mm/1024 pixels) (10 pixels) = 35 μm.

The autocorrelation intensity of the corresponding speckle patterns is shown in Fig. 1.4a and b. The figure shows that the aligned aperture is recognized from its speckle pattern (Fig. 1.4a) since it is different from the tilted aperture (Fig. 1.4b). Hence, any misalignment due to aperture tilting is recognized by referring to different speckle patterns.

The autocorrelation of the four equally spaced ellipses where the angle between each pair of ellipses is θ = 45° as shown in Fig. 1.5.

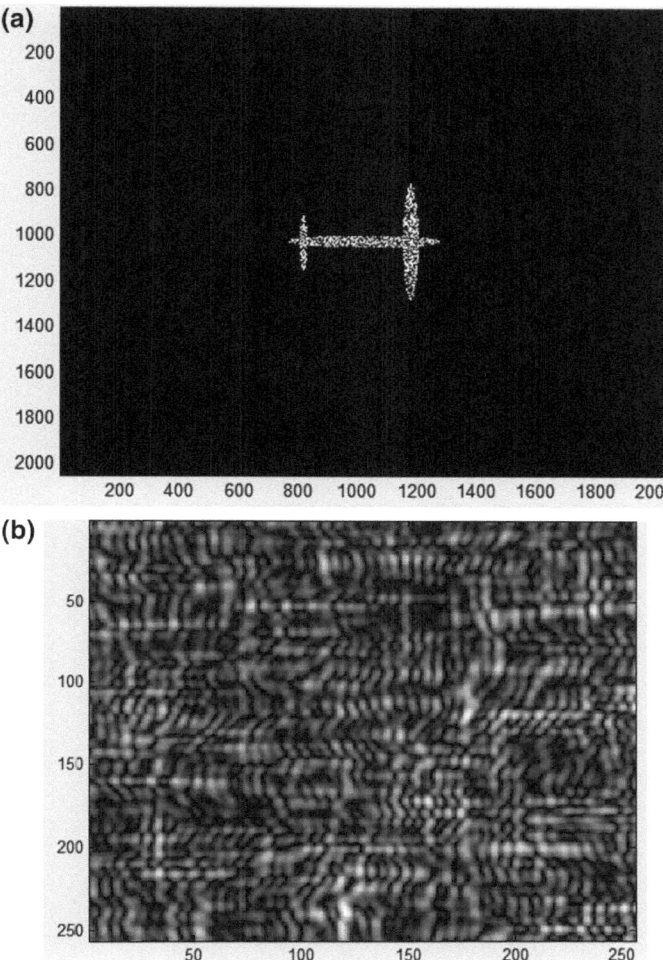

Fig. 1.2 a A pupil in the form of an airplane superimposed over the diffuser with matrix dimensions of 2048 × 2048 pixels, **b** Speckle elongation of dimensions 256 × 256 pixels corresponding to the airplane diffused pupil of 2048 × 2048 pixels

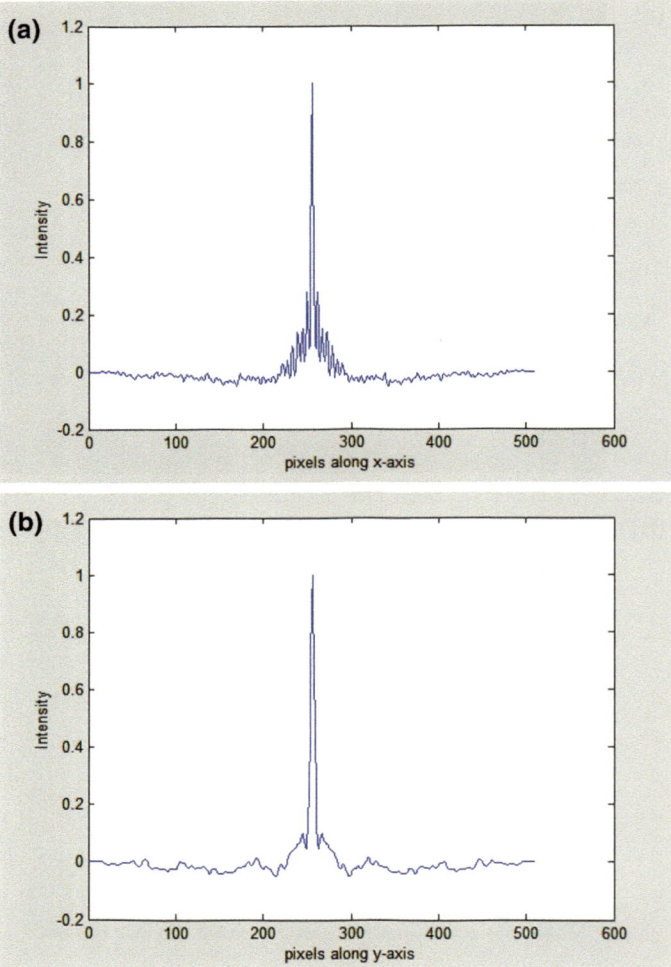

Fig. 1.3 a Autocorrelation intensity along the x-axis of the elongated speckles shown in Fig. 1.2b for the airplane pupil of 2048×2048 pixels. The average speckle size is equal to $\sigma_x = (4.5 \text{ mm}/512$ pixels) (5 pixels) $= 44\ \mu$m, **b** Autocorrelation intensity along the y-axis of the elongated speckles shown in Fig. 1.2b for an airplane pupil of 2048×2048 pixels. The average speckle size is equal to $\sigma_y = (3.6 \text{ mm}/512$ pixels) (8 pixels) $= 56\ \mu$m

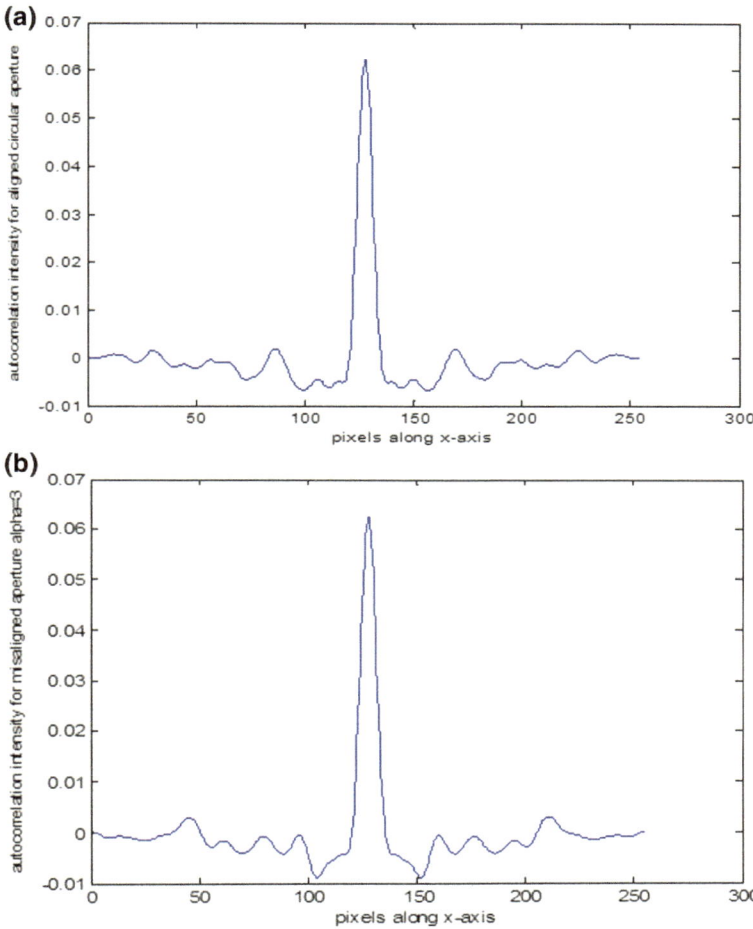

Fig. 1.4 a Autocorrelation intensity of the speckle pattern corresponding to the aligned circular aperture, **b** Autocorrelation intensity of the speckle pattern corresponding to a misaligned circular aperture with a tilting angle $\alpha = 3°$

Fig. 1.5 Autocorrelation of
the four crossed ellipses

1.4 Conclusion

First, for the elliptical apertures, that direction is easily recognized from the appearance of elongated speckles oriented normally to the major axis of the ellipse.

Second, for the two crossed orthogonal ellipses the speckles elongate in the two orthogonal directions as shown in the figures.

Third, for the other two models, namely the four crossed ellipses and the airplane model, the elongation is relatively complicated since each part gives rise to elongation normal for the elliptical part of interest.

The effect of aperture tilting was investigated for sharp elliptic apertures and compared with that for an aligned elliptic aperture. This allows the recognition of the aligned aperture compared with the misaligned tilted elliptic apertures. Additionally, three different speckle images obtained for circular aperture manipulation for comparison show a significant difference between the aligned aperture and the misaligned tilted circular apertures.

References

1. G.G. Mu et al., White-light image processing using oriented speckle-screen encoding. Opt. Lett. **10**, 375–377 (1985)
2. N. Barakat, A.M. Hamed et al., A photographic encoder applied to an optical processor using speckle techniques. J. Mod. Opt. **38**, 203–208 (1991)
3. A.M. Hamed, A.M. Discrimination, Between speckle images using deformed apertures. Opt. Eng. **56**, 1–7 (2011)
4. P. Lehmann, Surface roughness measurement by polychromatic speckle elongation. Appl. Opt. **36**, 2188–2197 (1997)

5. P. Lehmann, The aspect ratio of elongated polychromatic far-field speckles of continuous and discrete spectral distributions concerning surface roughness characterization. Appl. Opt. **41**, 2008–2014 (2002)
6. Z.H. Yuan et al., Comparison of surface roughness measurement in dichromatic speckle patterns with autocorrelation method. Adv. Mater. Res. **680**, 189–193 (2011)
7. A. Senthil Kumar, R.M. Vasu, Imaging with the oriented photographic diffuser. Instrumentation and Services Unit, Indian Institute of Science, Bengaluru (2002)
8. H. Lin, Speckle mechanism in holographic optical coherence imaging, Ph.D. dissertation from University of Missouri, pp. 57, 58 (2009)
9. H. Lin, P. Yu, Speckle mechanism in holographic optical imaging. Opt. Express **15**, 16322–16327 (2007)
10. J.W. Goodman, *Statistical Properties of Laser Speckle Patterns in Laser Speckle and Related Phenomena* (Springer-Verlag, New York, 1984)
11. Y. Piederriere et al., Particle aggregation monitoring by speckle size measurement, application to blood platelets aggregation. Opt. Express **12**, 4596–4601 (2004)
12. P. Lehmann, Surface roughness measurement based on the intensity correlation function of scattered light under speckle pattern illumination. Appl. Opt. **38**, 1144–1152 (1999)
13. R. Berlasso et al., Study of speckle size of light scattered from rough cylindrical surfaces. Appl. Opt. **34**, 5811–5819 (2000)
14. L.T. Alexander et al., Average speckle size as a function of intensity threshold level: comparison of experimental measurements with theory. Appl. Opt. **33**, 8240–8250 (1994)
15. A.M. Hamed, Recognition of direction of new apertures from the elongated speckle images: simulation. Opt. Photon. J. **3**, 250–258 (2013). https://doi.org/10.4236/opj.2013.33040

Chapter 2
Speckle Images Using Gaussian and Graded Index Apertures

2.1 Introduction

Recently, speckle formation was investigated using diffusers modulated by linear, conical, quadratic, obstructed circular, deformed kidney, or elliptical modulated apertures [1–4]. Using Fourier optics analysis, we showed that the speckle features for the above-mentioned apertures are dependent upon the aperture distribution. Others proposed a multiple exposure speckle-gram by using an optical system whose multiple aperture pupil changes between exposures [5]. The characteristics of speckle patterns generated through multi-aperture pupils were also theoretically analyzed based on Fourier techniques [6]. In another recent work, they analyzed speckle images generated when a diffuser was illuminated by coherent light and imaged by a lens with a pupil mask and multiple apertures forming a closed curve to obtain a clustered speckle structure [7]. The cluster structure results from the complex modulation produced inside each speckle, which is generated by multiple interferences of light through the apertures. When the apertures are uniformly distributed along a closed curve, the resulting image speckle cluster replicates the pupil aperture distribution. The authors in [7] showed from experimental and theoretical simulations that the cluster features are dependent on the aperture distribution and the size of the closed curve.

The speckle is considered an important topic for optical imaging of biomedical objects with irregular shapes, such as tumors and human skin.

The average speckle size of a speckle image is estimated by calculating the autocorrelation function of the digitized intensity speckle pattern. The autocorrelation function corresponds to the normalized autocorrelation function of the intensity, which has a zero base, and its full width at half maximum (FWHM) supplies a reasonable measurement of the average width of the speckle [8–10].

In the present chapter, modulated graded indices and truncated apertures are considered [11]. The speckle images of diffusers using these modulated apertures were examined. Hence, the autocorrelation function of the modulated speckle images

© The Author(s), under exclusive license to Springer Nature Switzerland AG 2024
A. Hamed, *Speckle Imaging Using Aperture Modulation*,
SpringerBriefs in Applied Sciences and Technology,
https://doi.org/10.1007/978-3-031-58300-1_2

is computed and plotted. The average speckle size for these modulated speckle images is computed from the autocorrelation of the speckle images.

2.2 Theoretical Analysis

2.2.1 Construction of the Gaussian, Graded, and Truncated Apertures

A numerical aperture of a definite number of steps (N) in the central black zone that increases gradually outwards, as shown in Fig. 2.1a, was investigated. This aperture is considered a graded index aperture.

This graded index aperture is represented mathematically as follows:

$$P_{\text{graded}}(x, y) = P_0 + \sum_{i=2}^{N} P_i(x, y) \tag{2.1}$$

P_0 is the central black zone of $i = 1$, α is a weighting factor with values of 0.2, 0.3…, 1, and N is the total number of graded steps.

The second aperture assumes successive black and transparent annuli in the central black zone, and a computerized view of this aperture is presented in Fig. 2.1b. This aperture is like the first aperture but with equal steps of a weighting factor $\alpha = 1$.

This aperture is like the graded index aperture with $\alpha = 1$ and is represented mathematically as follows:

$$P_{B/W}(x, y) = P_0 + \sum_{i=2}^{N} P_i(x, y) \tag{2.2}$$

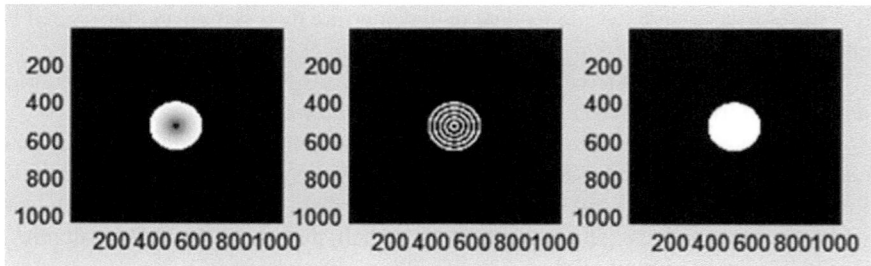

Fig. 2.1 First aperture from the left has a graded index aperture, the first aperture in the middle is a black and transparent concentric annular aperture, and the first aperture in the right is a circular aperture. All the apertures have dimensions of 1024 × 1024 pixels and a radius of $R = 128$ pixels

The third truncated aperture is represented as follows:

$$P_{\text{truncated}}(x, y) = \text{circ}(\rho, \theta) = 1; \; |\frac{\rho}{\rho_0}| \leq 1 \qquad (2.3)$$

$\rho = \left(x^2 + y^2\right)^{1/2}$. The letter ρ is the radial coordinate in the aperture plane.

Four different sections from the circular aperture were chosen as follows:

The first is a quarter of a circle in the range $0 \leq \theta \leq \pi/2$; the second is half of a circle, and its azimuthal range is $0 \leq \theta \leq \pi$; the third is three-quarters of a circle, and its azimuthal range is $0 \leq \theta \leq 3\pi/2$; and the fourth has two conjugate quarters or fan structures in the azimuthal range $0 \leq \theta \leq \pi/2$ and $\pi \leq \theta \leq 3\pi/2$.

A diffuser $d(x, y)$ considered a randomly distributed object is used to fabricate a speckle pattern in the case of the graded index apertures for a variable weighting factor α represented in Eq. (2.1) and the case of a step-index aperture of a weighting factor $\alpha = 1$ known as the black and transparent annulus Eq. (2.2), and the truncated apertures Eq. (2.3).

In the case of coherent uniform illumination, the transmitted complex amplitude becomes:

$$A(x, y) = P(x, y) \cdot d(x, y) \qquad (2.4)$$

where $p(x, y) = P_{\text{graded}}(x, y)$; for the graded index aperture, $p(x, y) = P_{B/W}(x, y)$; for the B/W step-index aperture, and $p(x, y) = P_{\text{truncated}}(x, y)$; for truncated apertures.

The speckle pattern was obtained in the focal plane of the imaging lens using coherent laser illumination by running the Fourier transform on Eq. (2.4) to obtain:

$$B(u, v) = \text{F.T.}\{A(x, y)\} = \text{F.T.}\{P(x, y) \cdot d(x, y)\} = h(u, v) * D(u, v) \qquad (2.5)$$

The symbol (*) is used for the convolution operation, $h(u, v) = \text{F.T.}\{p(x, y)\}$ and $D(u, v) = \text{F.T.}\{d(x, y)\}$ is the conventional speckle image in the case of a uniform circular aperture where $p(x, y) = 1$ inside the aperture and is zero outside the aperture.

2.2.2 Computation of the Autocorrelation of the Modulated Speckles

The modulated speckles are constructed from the operation of the Fourier transform of the multiplication of the diffuser and the modulating aperture, represented in Eq. (2.5). The autocorrelation of the speckle image was obtained using MATLAB code. The autocorrelation function of Eq. (2.5) is written as follows:

$$c(u, v) = B(u, v) \otimes \ \ B(u, v) \tag{2.6}$$

The symbol $(\otimes)$ is used for the autocorrelation operation of two similar speckle images Eq (2.6).

If the complex amplitude of the modulated speckle image given by Eq. (2.5) is rewritten as follows:

$$D_{\text{modulating}} = h(u, v) \otimes \ \ D(u, v); \ \ \text{where} \ \ B(u, v) = D_{\text{modulating}} \tag{2.7}$$

Hence, the autocorrelation product ran on two symmetric modulated apertures as shown in Eq. (2.8).

$$c(u, v) = D_{\text{modulating}} \otimes \ \ D_{\text{modulating}} \tag{2.8}$$

2.3 Results and Discussion

Three different apertures are numerically constructed, as shown in Fig. 2.1. The first aperture from the left has a graded index aperture, the middle has a black and transparent concentric annular aperture, and the right has a circular aperture. All the plotted apertures have dimensions of 1024×1024 pixels and a radius of $R = 128$ pixels.

Another distinct set of truncated apertures is plotted as in Fig. 2.2.

From the left, the first has a three quarters of a circular aperture, the second is a half circular aperture, the three has a quarter (2.3), and the fourth has two symmetric quarters in the form of a fan aperture.

The speckle images for diffusers modulated by the described apertures shown in Figs. 2.1 and 2.2 were obtained by running the Fourier transform of the multiplication product of the diffuser and the aperture. Hence, the modulated speckle images shown in Figs. 2.3 and 2.4 are the convolution products of the ordinary speckle image and the point spread function (PSF) of the defined aperture. The three speckle images shown in Fig. 2.3 are completely different since the PSF is different for each aperture. Additionally, the speckle images shown in Fig. 2.4a–d are dependent upon the PSF of the truncated apertures.

The five profile line shapes of the three speckle images with the arrangement shown in Fig. 2.3 at lines 50, 110, 128, 160, and 190 are shown in Fig. 2.5a–e. For each graph, three curves are plotted, the upper curve for the speckle corresponds to the graded index aperture, the middle curve corresponds to the step-index B/W aperture, and the lower curve corresponds to a circular uniform aperture. The three profiles for each of the five graphs are completely different since the apertures are different which is in good agreement with the graphs shown in Fig. 2.3. This is attributed to the convolution product of the ordinary speckle in the case of the diffuser and the PSF of the different apertures.

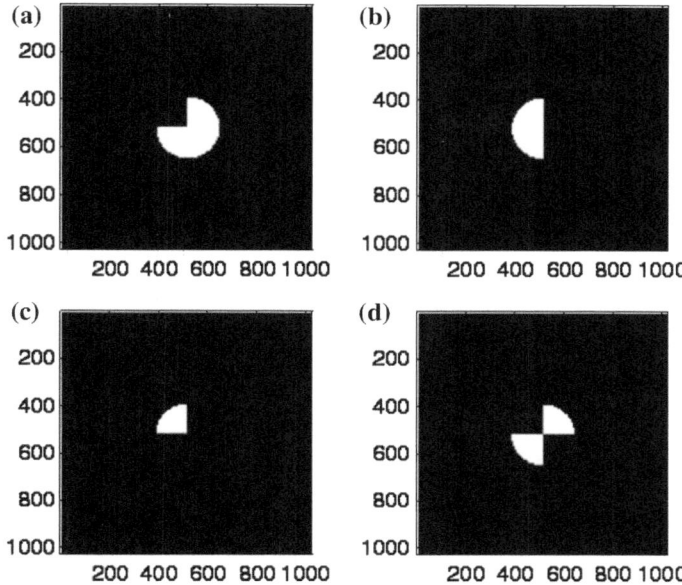

Fig. 2.2 Four different truncated apertures, where **a** three-quarters aperture, **b** semicircular aperture, **c** one-quarter circular aperture, and **d** two conjugate quarters in the form of a fan

Fig. 2.3 From the left, speckle images corresponding to the diffuser provided with the apertures are shown in Fig. 2.1

A separate profile line shape of the speckle images corresponding to the diffuser provided with the graded index aperture is shown in Fig. 2.6a. The profile line shapes of the speckle images obtained using the step-index black and white concentric annular aperture are shown in Fig. 2.6b. The comparative line shapes of the speckle images corresponding to the diffuser provided with the circular uniform aperture are shown in Fig. 2.6c. All the curves corresponding to the three separate graphs starting from the upper plot are plotted at lines 50, 100, 150, and 200.

A uniform circular aperture with a radius = 64 pixels on a matrix with dimensions of 2048 × 2048 pixels is plotted on the left, while on the right is the speckle pattern

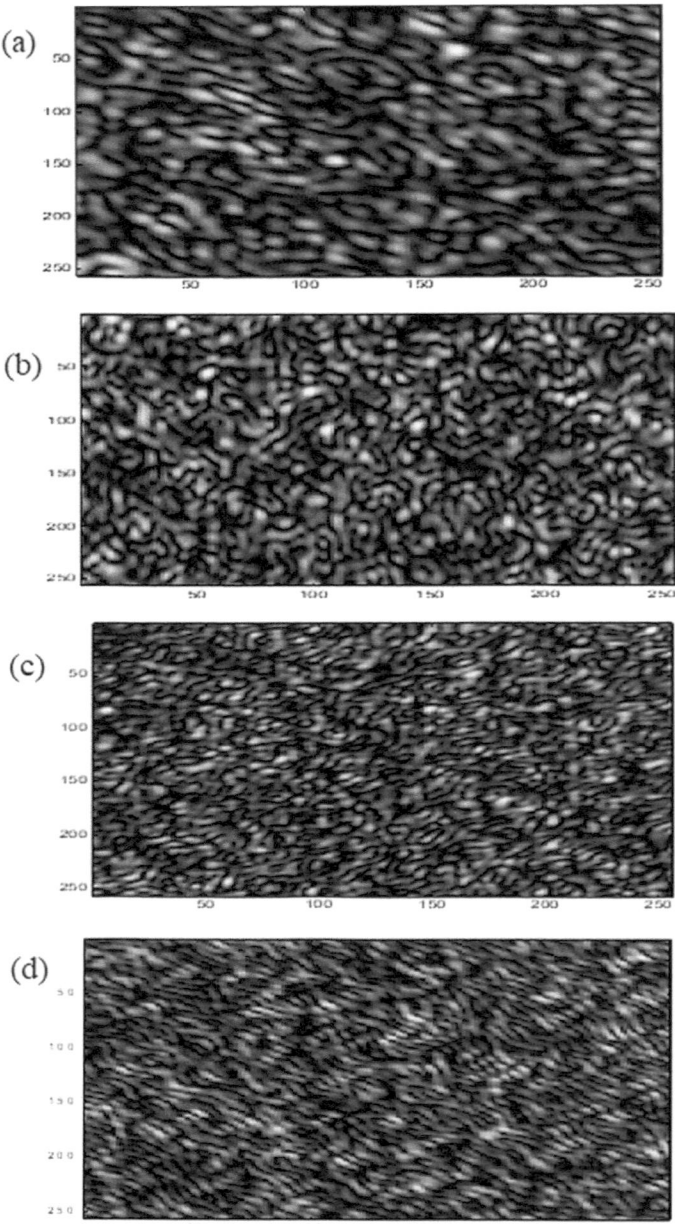

Fig. 2.4 Speckle pattern for the following apertures: **a** a quarter circular aperture, **b** a half circular aperture, **c** a three-quarter circular aperture, and **d** a fan aperture

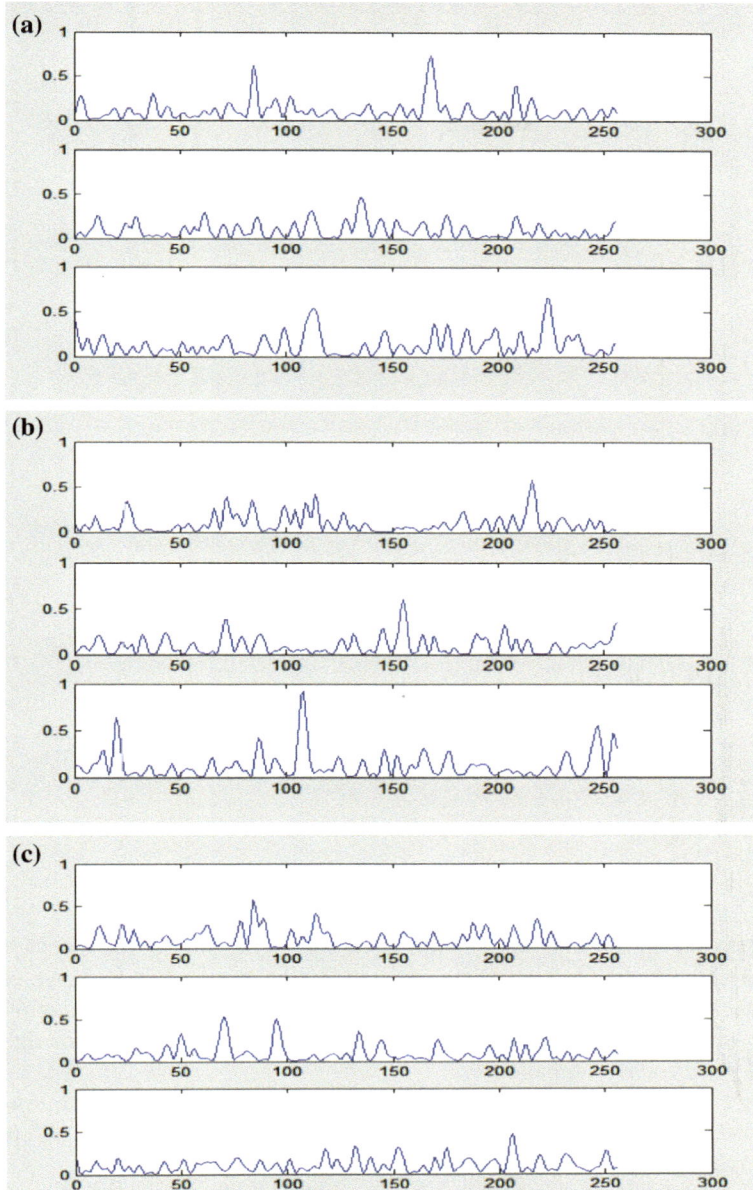

Fig. 2.5 The five profile line shapes of the three speckle images with the arrangement shown in Fig. 2.3 at lines 50, 110, 128, 160, and 190 pixels

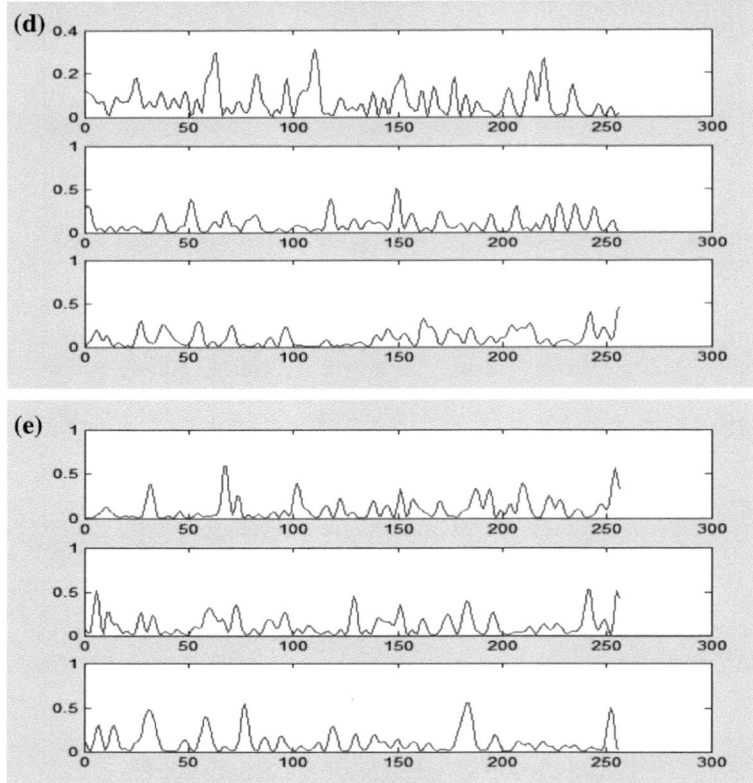

Fig. 2.5 (continued)

of the diffuser provided with this aperture which has dimensions of 256×256 pixels as shown in Fig. 2.7. The field of view is 4.5 mm $\times$ 3.6 mm for the speckle image.

The autocorrelation function corresponds to the above speckle image obtained from the same diffuser provided with a circular aperture with a radius of 64 pixels as shown in Fig. 2.8. On the left is the peak along the x-axis, while on the right is the peak for the y-axis summation.

The FWHM of the speckle size corresponding to the 20 pixels shown on the autocorrelation peak, along the x-axis and y-axis, is equal to $\sigma_x = (1/2)$ (4.5 mm/512 pixels) (20 pixels) $= 88$ µm.

The FWHM of the speckle size corresponding to the 20 pixels shown on the autocorrelation peak along the y-axis is equal to $\sigma_y = (1/2)$ (3.6 mm/512 pixels) (20 pixels) $= 70$ µm.

The autocorrelation images of the truncated apertures are plotted in Fig. 2.9.

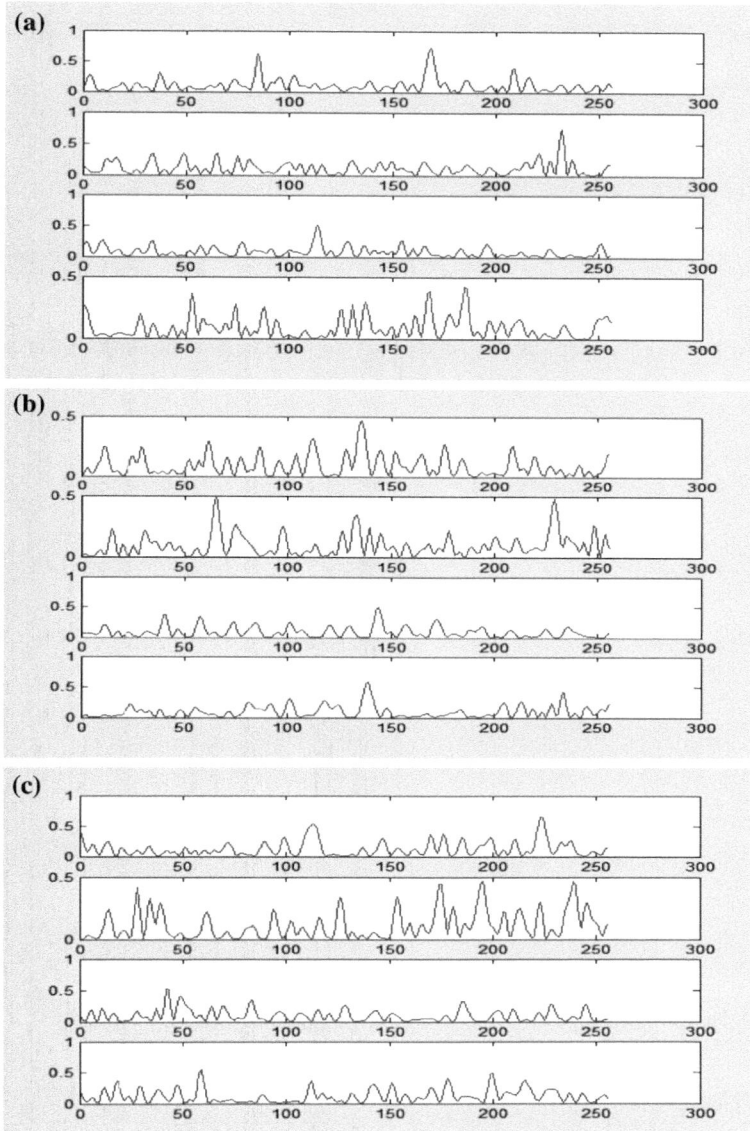

Fig. 2.6 A separate profile line shape of the speckle images corresponding to the diffuser provided with the graded index aperture is shown in Fig. 2.6a at lines 50, 100, 150, and 200 pixels. Similar plots are given for the step-index aperture Fig. 2.6b and the uniform circular aperture Fig. 2.6c

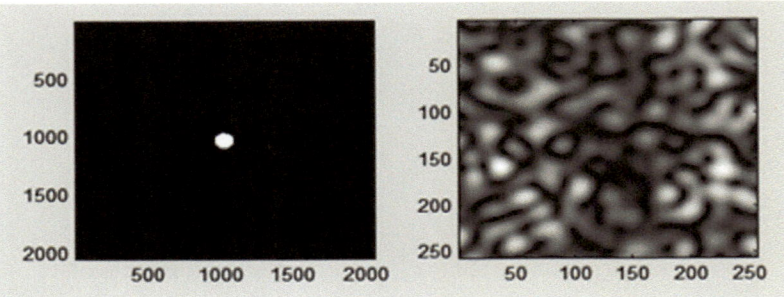

Fig. 2.7   A small circular aperture of radius = 64 pixels and the corresponding speckle image

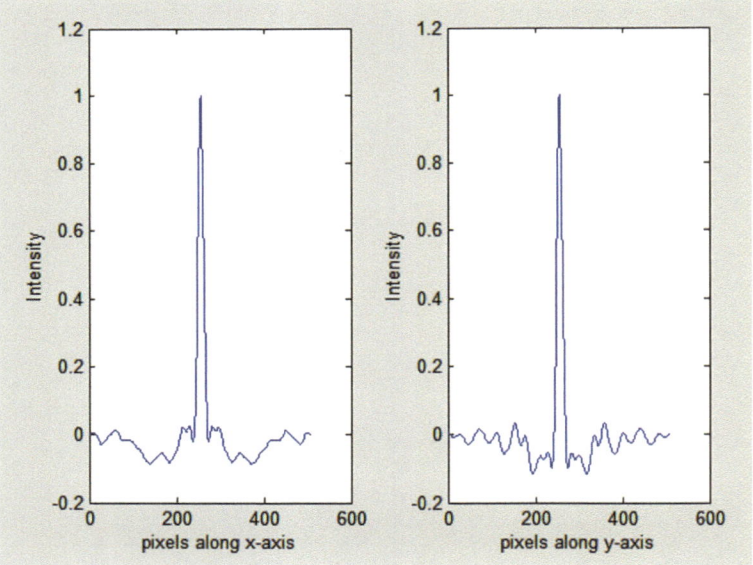

Fig. 2.8   Autocorrelation profiles of speckles using a circular aperture

2.4   Conclusion

The speckle images obtained for diffusers with graded indices and truncated apertures are different from those obtained with uniform circular apertures. The autocorrelation of the speckle images was plotted, and the FWHM of the speckle images was computed. Additionally, speckle images were obtained for truncated apertures and the differences were noticeable. It was concluded that the different speckle images are dependent upon the point spread function of the examined aperture. Consequently, the modulated speckle image is the convolution product of the ordinary speckle image

Fig. 2.9 Autocorrelation images of the truncated apertures. **a** For the quarter, **b** for half of the circle, **c** for three-quarters of the circle, and **d** for the fan aperture

obtained in the case of a uniform circular aperture and the point spread function of the modulating aperture.

References

1. A.M. Hamed, Numerical speckle images using modulated linear apertures: simulation. J. Mod. Opt. **56**, 1174–1181 (2009)
2. A.M. Hamed, Formation of speckle images using circular obstruction. J. Mod. Opt. **56**, 1633–1642 (2009)
3. A.M. Hamed, Discrimination between speckle images using diffusers modulated by some apertures: simulations. Opt. Eng. **50**, 1–7 (2011)
4. A.M. Hamed, Computer generated quadratic and higher order apertures and its application on numerical speckle images. Opt. Photon. J. **1**, 41–52 (2011)
5. M. Tebaldi, L.A. Toro, New multiple aperture arrangements for speckle photography. Opt. Commun. **182**, 95–105 (2000)
6. L.A. Toro, M. Tebaldi, Properties of speckle patterns generated through multi-aperture pupils. Opt. Commun. **192**, 37–47 (2001)
7. E. Mosso, M. Tebaldi et al., Cluster speckle structures through multiple apertures forming a closed curve. Opt. Commun. **283**, 1285–1290 (2009)
8. H. Lin, Thesis "Speckle mechanism in holographic optical coherent imaging," presented to the University of Missouri (2009)
9. H. Lin, Speckle mechanism in holographic optical imaging. Opt. Exp. **15**, 16322–16327 (2007)
10. J.W. Goodman, *Statistical Properties of Laser Speckle Patterns* (Springer, New York, 1984)
11. A.M. Hamed, Study of graded index and truncated apertures using speckle images. Precision Instrum. Mech. PIM **3**, 144–152 (2014)

Chapter 3
The Contrast of Speckle Images Using the Voigt Distribution

3.1 Introduction

Laser speckle contrast imaging (LSCI), first introduced in the early 1980s [1], is a robust, low-cost method allowing noncontact, full-field, and real-time flow system mapping.

The speckle pattern fluctuates if the illuminated area holds moving particles such as moving red cells. By integrating the intensity fluctuations of the speckle pattern over a finite time, we obtain information about the motion of the scattering particles.

We studied the probability density function and the average contrast of laser speckles in the diffraction field with the assumption of Gaussian statistics for forming speckles. The general equation of average contrast V and the probability density function P of the produced speckles are assumed [2, 3].

In 1997, they investigated velocity maps of blood flow in capillaries using a laser speckle contrast technique based on first-order statistics [4–6]. Laser speckle contrast analysis (LASCA) can produce 2D contrast maps indicating changes in velocity.

Hamed [7–12] published research on the contrast of speckle images using modulated apertures and compared it with that obtained in the case of a uniform circular aperture for illumination. He showed that the point spread function (PSF) improved in the modulated gaps compared to the uniform circular aperture for a specific N.A. and wavelength. Hence, the resolution of speckle images in the case of a modulated aperture is better than that obtained for a uniform circular aperture because of the improvement in PSF.

In [13], we investigated speckle imaging of an annular Hermite Gaussian laser beam. The results from the distribution permit computation of the speckle size from the full width at half maximum (FWHM) corresponding to the PSF of the Hermite Gaussian annular aperture.

In [14–18], blood flow was recently investigated using speckle techniques. These studies used capillary blood flow monitoring via laser speckle contrast analysis [14].

© The Author(s), under exclusive license to Springer Nature Switzerland AG 2024
A. Hamed, *Speckle Imaging Using Aperture Modulation*,
SpringerBriefs in Applied Sciences and Technology,
https://doi.org/10.1007/978-3-031-58300-1_3

Optic nerve head blood flow, as measured by laser speckle flow graphics, is significantly reduced in perimetric glaucoma [15]. An experimental study on correlation and contrast showed the effect of static scatterers in laser speckle contrast imaging [16]. In [17], blood perfusion in human eyelid skin flaps was examined via laser speckle contrast imaging.

In [1, 5], the authors computed the fluid velocity from the relation between the speckle contrast and the field autocorrelation.

In this chapter, I present an analysis of the image formation of speckles using a laser sheet for fluid illumination based on convolution and Fourier techniques. I considered the Voigt distribution in the time domain to obtain a novel formula for the contrast of speckle images.

3.2 Theoretical Analysis

The complex amplitude corresponding to the laser beam propagation in the zero-order mode is assumed to follow a Gaussian distribution as follows:

$$A(r, z) = \frac{w_0}{w(z)} \exp\left(-\frac{r^2}{w(z)^2}\right), \tag{3.1}$$

where $w(z)$ is the waist of the beam at propagation distance z. The relation between $W(z)$ and the waist of the beam w_0 at $z = 0$ is the following:

$$w^2(z) = w_0^2\left[1 + (z/z_0)^2\right], \quad \text{and} \quad z_0 = \pi w_0^2/\lambda$$

Referring to Eq. (3.1) and assuming $z = 0$, where $w(z) = w_0 = w$, we represent the laser sheet propagation as follows:

$$A(x, y = c, z = 0) = \exp\left(-\frac{x^2}{w^2}\right) \cdot \text{rect}(x, y) \tag{3.2}$$

The rectangular function is the envelope of the laser sheet and is represented as follows:

$$\text{rect}(x, y) = 1 \quad \text{for} \quad \left|\frac{x}{x_s}\right| \leq 1 \quad \text{and} \quad y = \text{const.} \tag{3.3}$$

The width of the laser sheet at constant height y is x_s.

A fluid that moves steadily in a specific direction is represented at a particular section z in discrete form as follows:

$$g(x, y; z = \text{const.}) = \sum_{m=0}^{M}\sum_{n=0}^{N} g(x - m\Delta x, y - n\Delta y), \quad v = 0 \tag{3.4}$$

For fluid moving with velocity v in a direction parallel to the sheet width x, we rewrite Eq. (3.4) as follows:

$$g(x, y; z = \text{const.}) = \sum_{t=0}^{T} \sum_{m=0}^{M} \sum_{n=0}^{N} g(x - m\Delta x - v\Delta t, \; y - n\Delta y) \qquad (3.5)$$

T is the total exposure time for speckle images.

Using Eqs. (3.2) and (3.5), the complex amplitude of the laser sheet traversing the fluid can be represented as follows:

$$B(x, y = c) = A(x, y = c) \cdot g(x, y) \qquad (3.6)$$

In the focal plane of a lens or at a great distance between the fluid and the CCD camera, allowing us to perform Fraunhofer diffraction, we run the Fourier transform on Eq. (3.6) written in integral form as follows:

$$\tilde{B}(u, v) = \text{F.T.}\{B(x, y)\} = \int_{-\infty}^{\infty} \int B(x, y) \exp\left\{ -\frac{j2\pi}{\lambda f}(xu + yv) \right\} \mathrm{d}x \; \mathrm{d}y \qquad (3.7)$$

Symbolically, referring to Eq. (3.6), we write:

$$\tilde{B}(u, v) = \text{F.T.}\{A(x, y = c) \cdot g(x, y)\} \qquad (3.8)$$

By utilizing the properties of the Fourier transform and convolution operation, we obtain the following integral:

$$\tilde{B}(u, v) = \int_{-\infty}^{\infty} \int \tilde{A}(u', v') \; \tilde{g}(u - u', v - v') \mathrm{d}u' \; \mathrm{d}v' \qquad (3.9)$$

Symbolically, we represent Eq. (3.9) as follows:

$$\tilde{B}(u, v) = \tilde{A}(u, v) \otimes \tilde{g}(u, v) \qquad (3.10)$$

$\otimes$: symbol for convolution operation.

$$\tilde{A}(u, v) = \text{F.T.}\{A(x, y)\} \text{ and } \tilde{g}(u, v) = \text{F.T.}\{g(x, y)\}$$

Running the F.T. upon Eq. (3.8) gives $\tilde{B}(u, v)$. This phenomenon occurs in the focal plane f or at a great distance z compared with the laser sheet width.

$$\tilde{B}(u, v) = \frac{\sin(\pi u)}{\pi u} \otimes \exp(-w^2 u^2) \otimes \sum_{t=0}^{T} \sum_{m=0}^{M} \sum_{n=0}^{N} \tilde{g}(u - m/\Delta x - v/\Delta t, v - n/\Delta y)$$

(3.11)

Equation (3.11) is the complex amplitude corresponding to the speckle pattern produced by applying the laser sheet to the fluid in a tube considered an object.

The image intensity corresponds to the recorded speckle image from the equation's modulus squared (3.11). We write as follows:

$$I(u, v) = \left| \frac{\sin(\pi u)}{\pi u} \otimes \exp(-w^2 u^2) \otimes \sum_{t=0}^{T} \sum_{m=0}^{M} \sum_{n=0}^{N} \tilde{g}\left(u - \frac{m}{\Delta x} - \frac{v}{\Delta t}, v - \frac{n}{\Delta y}\right) \right|^2$$

(3.12)

In a case where the laser sheet width is exceedingly small approximated by a Dirac delta function, the speckle image in Eq. (3.12) is written as follows:

$$I(u, v) = \left| \exp(-w^2 u^2) \otimes \sum_{t=0}^{T} \sum_{m=0}^{M} \sum_{n=0}^{N} \tilde{g}(u - m/\Delta x - v/\Delta t, v - n/\Delta y) \right|^2$$

(3.13)

Consequently, $\exp(-w^2 u^2)$ the point spread function (PSF) or the amplitude impulse response affects the speckle distribution via a Gaussian shape. We produced randomness in speckle images from the randomness of fluid flow convoluted with the Gaussian function. Moreover, the PSF corresponds to the circular aperture represented by the known Airy disk.

We computed the contrast of the speckle image from the following equation:

$$K = \frac{\sigma}{\langle I \rangle}$$

(3.14)

σ is the root mean square (RMS) value corresponding to intensity I, and $<I>$ is the mean intensity corresponding to the speckle image.

3.3 Relation Between the Contrast and the Speckle Field Autocorrelation

3.3.1 Case of Lorentzian Distribution

The relation between the contrast (K) and the speckle field autocorrelation $g_1(\tau)$ is written as follows [1, 14, 19] and represented by Eq. (3.15), where the field obeys the Lorentzian velocity distribution:

$$K = \sqrt{\frac{1}{T} \int_0^T |g_1(\tau)|^2 d} = \left[\left(\frac{\tau_c}{2T} \right) \left\{ 1 - \exp\left(-\frac{2T}{\tau_c} \right) \right\} \right]^{1/2} \tag{3.15}$$

where we represent the field autocorrelation as follows: $g_1(\tau) = \exp\left(-\frac{\tau}{\tau_c} \right)$.

Considering the autocorrelation time of the CCD camera represented by the triangular function, we obtain the contrast K, which is computed as follows [20]:

$$K = \sqrt{\frac{1}{T} \int_0^T \left(1 - \frac{T}{\tau_c} \right) |g_1(\tau)|^2 d\tau} \tag{3.16}$$

Solving for K, we obtain:

$$K = \left[\left(\frac{\tau_c}{T} \right) + \frac{1}{2} \left(\frac{\tau_c}{T} \right)^2 \left\{ 1 - \exp\left(-\frac{2T}{\tau_c} \right) \right\} \right]^{1/2} \tag{3.17}$$

The decorrelation velocity v_c is related to the wavelength λ and the decorrelation time τ_c, as shown in Eqs. (3.15) and (3.17), respectively [14]:

$$v_c = \frac{\lambda}{2_c} \tag{3.18}$$

Using Eqs. (3.14) and (3.15), we computed the decorrelation time τ_c from Eq. (3.16). Hence, we deduced the fluid velocity by knowing the contrast values and the exposure time T in seconds.

3.3.2 Gaussian Distribution Case

When the field obeys a Gaussian velocity distribution, we obtain the relation between the contrast K and the autocorrelation field $g_1(\tau)$. It has the following formula ref. [18, 21] and is represented by Eq. (3.19):

$$K = \left[\frac{\sqrt{\pi}}{2} \left(\frac{\tau_c}{T} \right) \mathrm{erf} \left(\frac{T}{\tau_c} \right) \right]^{1/2} \tag{3.19}$$

3.3.3   Case of the Voigt Distribution

We know that the line shape of the laser beam considering natural and Doppler broadening is computed from the following convolution product:

$$G(v) = \int_{-\infty}^{\infty} F(v') H(v - v') dv' \tag{3.20}$$

$$F(v) = \alpha \exp\left[-\beta (v - v_0)^2 \right] \tag{3.21}$$

α and β are constants.

$$H(v) = \frac{2t_{\mathrm{sp}}}{\pi} \frac{1}{1 + 16\pi^2 t_{\mathrm{sp}}^2 (v - v_0)^2} \tag{3.22}$$

$H(v)$ is the Lorentzian distribution for natural broadening in the frequency domain, and t_{sp} is the spontaneous time constant.

Equation (3.20) is written in symbolic form as follows:

$$G(v) = F(v) \otimes H(v) \tag{3.23}$$

We apply the Fourier transform to Eq. (3.23) and make use of the convolution operation we easily obtain the following result:

$$g(t) = f(t) \cdot h(t) \tag{3.24}$$

$g(t) = FT\{G(v)\}$ and similar expressions are given for $f(t)$ and $h(t)$.

We use the expressions for the Lorentzian and Gaussian line shapes represented in Eqs. (3.21, 3.22) we can write the electric field in the time domain as follows:

$$g'(t) = g_G'(t) \cdot g_L'(t) \tag{3.25}$$

The autocorrelation for both the Lorentzian and Gaussian beams is represented by the following symbolic relations:

$$g_G(t) = g_G'(t) \otimes g_G'(t) = \exp\left(-\frac{\tau}{2\tau_c^2} \right)^2 \tag{3.26}$$

$$g_L(t) = g'_L(t) \otimes g'_L(t) = \exp\left(-\frac{\tau}{\tau_c}\right) \tag{3.27}$$

Finally, we can write the autocorrelation of the electric field in the case of the Voigt distribution in the time domain as follows:

$$g(t) = \exp\left(-\frac{\tau}{2\tau_c^2}\right)^2 \cdot \exp\left(-\frac{\tau}{c}\right) \tag{3.28}$$

We make use of the relation between the contrast K and the field autocorrelation defined in Eq. (3.15) and substitute the field autocorrelation $g(t)$ with Eq. (3.28) we write the following:

$$K^2 = \frac{\sigma^2}{\langle I \rangle^2} = \frac{1}{T}\int_0^T |g_1(\tau)|^2 d\tau = \frac{1}{T}\int_0^T \left|\exp\left(-\frac{\tau^2}{2\tau_c^2}\right) \cdot \exp\left(-\frac{\tau}{\tau_c}\right)\right|^2 d\tau \tag{3.29}$$

We rewrite Eq. (3.29) as follows:

$$K^2 = \frac{\tau_c}{T}\exp(1)\int_0^T \exp\left[-\left(\frac{\tau}{\tau_c}+1\right)^2\right]d\tau \tag{3.30}$$

Solving the above integral, we finally obtain:

$$K^2 = \frac{\sqrt{\pi}}{2}\left(\frac{\tau_c}{T}\right)\exp(1) \cdot \mathrm{erf}\left(\frac{T}{\tau_c}+1\right) \tag{3.31}$$

The general expression of the Gauss error function is represented as:

$$\mathrm{erf}(T) = \frac{2}{\sqrt{\pi}}\int_0^T \exp(-t^2)dt \tag{3.32}$$

Hence, in the case of the Voigt distribution, we get shifted erf like in the case of the Gaussian distribution due to the presence of the Lorentzian distribution in the time domain. The contrast ratio between the Gaussian distribution and Voigt distribution is computed as follows:

$$\frac{K_G}{K} = 0.6065\sqrt{\frac{\mathrm{erf}\left(\frac{T}{\tau_c}\right)}{\mathrm{erf}\left(\frac{T}{\tau_c}+1\right)}} \tag{3.33}$$

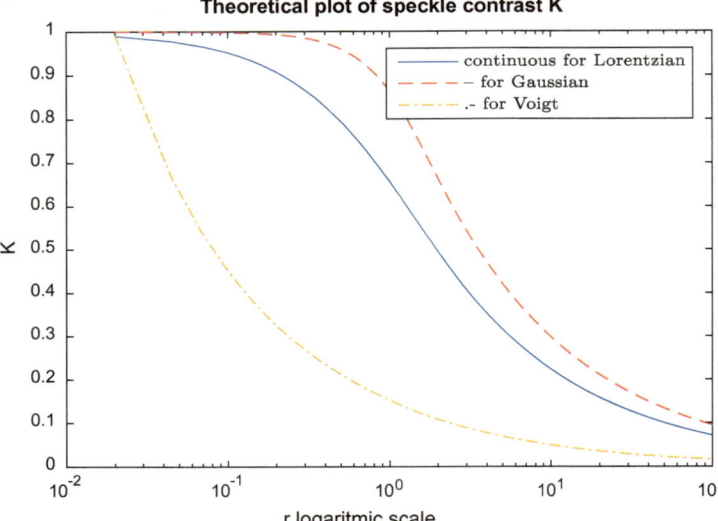

Fig. 3.1 Theoretical curves for the contrast (K) versus the ratio r between the exposure time (T) and the autocorrelation time for a water flow rate of 50 s

Consequently, the contrast assuming a Gaussian distribution is <60% of the contrast in the case of the Voigt distribution. We compare the contrasts for the Lorentzian, Gaussian, and Voigt distributions as shown in the plot.

3.4   Results and Discussions

The theoretical curves for the speckle contrast (K) versus the ratio r between the exposure time (T) and the autocorrelation time are plotted in Fig. 3.1. Three curves are shown, and the continuous curve corresponds to the Lorentzian distribution shown in Eq. (3.17) where we considered the autocorrelation time of the CCD camera represented by the triangular function for the rectangular window. In addition, the discontinuous curve corresponds to the field obeying the Gaussian velocity distribution shown in Eq. (3.19). The line shape of the laser beam is computed from the convolution product corresponding to the natural and Doppler broadening and is Fourier transformed to obtain the speckle contrast K in Eq. (3.31). Hence, we obtain the dotted curve that corresponds to the Voigt distribution.

References

1. A.F. Fercher, J.D. Briers, Flow visualization with single-exposure speckle photography. Opt. Commun.Commun. **37**(5), 326–330 (1981)
2. J.C. Dainty (ed.), *Laser Speckle and Related Phenomena* (Springer, Berlin, 1975)
3. J. Ohtsubo, T. Asakura, Statistical properties of laser speckle produced in the diffraction field. Appl. Opt. **16**(6), 1742–1753 (1977)
4. G.J. Richards, J.D. Briers, Laser speckle contrast analysis (LASCA): a technique for measuring capillary blood flow using the first order statistics of laser speckle patterns, in *IEE Colloquium on Biomedical Applications of Photonics (Digest No. 1997/124)* (IET, 1997), pp. 11–1
5. H. Cheng, Q.T. Duong, Simplified laser-speckle-imaging analysis and its application to retinal blood flow imaging. Opt. Lett. **32**(15), 2188–2190 (2007)
6. D.D. Duncan, S.J. Kirkpatrick, Algorithms for simulation of speckle (laser and otherwise), 6855, 685505-685505-8 (2008). doi:https://doi.org/10.1117/12.760518
7. A.M. Hamed, Formation of speckle images formed for diffusers illuminated by modulated apertures (circular obstruction). J. Mod. Opt. **56**(15), 1633–1642 (2009). https://doi.org/10.1080/09500340903277792
8. A.M. Hamed, Numerical speckle images formed by diffusers using modulated conical and linear apertures. J. Mod. Opt.Mod. Opt. **56**, 1174–1181 (2009). https://doi.org/10.1080/09500340902985379
9. A.M. Hamed, Discrimination between speckle images using diffusers modulated by some deformed apertures: simulation. J. Opt. Eng. **50**, 1–7 (2011). https://doi.org/10.1117/1.3530085
10. A.M. Hamed, Study of the graded index and truncated apertures using speckle images. Precision Instrum. Mech. PIM **3**, 144–152 (2014)
11. A.M. Hamed, Image processing of Ramses II statue using speckle photography modulated by a new Hamming-Linear aperture. Pram. J. Phys. **94**, 126 (2020)
12. A.M. Hamed, Contrast of laser speckle images using some modulated apertures. Pram. J. Phys. **95**, 122 (2021)
13. A.M. Hamed, Speckle imaging of annular Hermite Gaussian laser beam. Pram. J. Phys. **95**, 202 (2021)
14. J.D. Briers, G.J. Richards, X.W. He, Capillary blood flow monitoring using laser speckle contrast analysis (LASCA). J. Biomed. Opt. **4**(1), 164–176 (1999)
15. Y. Shiga, H. Kunikata et al., Optic nerve head blood flow, as measured by laser speckle flow graphic, significantly reduced in perimetric glaucoma. Curr. Eye Res.. Eye Res. **41**(11), 1447–1453 (2016)
16. P.G. Vaz, A.H. Humeau, S. Figueiras et al., Effect of static scatterers in laser speckle contrast imaging: an experimental study on correlation and contrast. Phys. Med. Biol. **63**(1), 015024 (2017)
17. C.D. Nguyen, J. Hult et al., Blood perfusion in human eyelid skin flaps examined by laser speckle contrast imaging—importance of flap length and the use of diathermy. Ophthalmic Plast. Reconstr. Surg.Plast. Reconstr. Surg. **34**(4), 361–365 (2018)
18. J.W. Goodman, *Speckle Phenomena in Optics: Theory and Applications* (Greenwood Village: Roberts & Company, 2006)
19. J.D. Briers, S. Webster, Quasi-real-time digital version of single-exposure speckle photography for full-field monitoring of velocity or flow fields. Opt. Commun.Commun. **116**, 36–42 (1995)
20. Z. Wang, S. Hughes et al., Theoretical and experimental optimization of laser speckle contrast imaging for high specificity to brain microcirculation. J. Cereb. Blood Flow Metab.Cereb. Blood Flow Metab. **27**, 258–269 (2007)
21. A.M. Hamed, Recognition of Direction of New Apertures from the Elongated Speckle Images: Simulation. Opt. Photon. J. **3**, 250–258 (2013). https://doi.org/10.4236/opj.2013.3304

Chapter 4
Contrast of Speckle Images Using Modulated Apertures

4.1 Introduction

Since the advent of lasers speckle patterns have been obtained from scattered and reflected light from coherent laser beams, and the statistical properties of speckles have been investigated [1–5]. Fourier optics techniques produce speckle images from the convolution product of the FFT of the diffuser and the point spread function (PSF) corresponding to the linear aperture placed in the diffuser plane [6]. Modulated speckle images by linear, quadratic, and other apertures were investigated in [7–9]. Laser speckle contrast imaging (LSCI) was first investigated in [10] where several workers focused on the features of dynamic speckles [11–14]. In a recent review of LSCI outlined in [15–17] theories relating speckle contrast to particle speed were given. The influence of Gaussian or Lorentzian velocity distributions on the contrast level was investigated in [18]. Coherence domain optical methods and optical coherence tomography in biomedicine were presented in [19]. Recently, the joint effect of scattering and absorption on laser speckle contrast imaging was investigated in [20] while a quantitative model of diffuse speckle contrast analysis for flow measurement was outlined in [21]. In addition, contrast analysis signals with absolute blood flow are given [22]. Image processing of the Ramses II statue using speckle photography modulated by a new Hamming linear aperture was investigated in [23].

In this chapter, laser speckle images are investigated using linear, quadratic, and Hamming apertures. The speckle contrast versus the speckle size and aperture was computed and discussed [24].

© The Author(s), under exclusive license to Springer Nature Switzerland AG 2024   35
A. Hamed, *Speckle Imaging Using Aperture Modulation*,
SpringerBriefs in Applied Sciences and Technology,
https://doi.org/10.1007/978-3-031-58300-1_4

4.2 Theoretical Analysis

A subjective laser speckle pattern is formed in the Fourier plane of a collimating lens L when a collimated laser beam is incident upon a diffuser obstructed by a linear aperture in a circular mask.

Hence, we write the following for the transmittance complex amplitude from the diffuser superposed over the linear aperture of radius r_0 as follows:

$$A(x, y) = d(x, y) \cdot P(x, y) \tag{4.1}$$

where $d(x, y) = a(x, y)\exp[2\pi i \cdot \text{rand}(x, y)]$ represents the complex amplitude of the diffuser, $a(x, y)$ is its amplitude, and $\exp[2\pi i \cdot \text{rand}(x, y)]$ is its phase. $P(r) = \frac{r}{r_0}; \left|\frac{r}{r_0}\right| \leq 1$ is the linear amplitude transmittance from the circular aperture where r is the radial coordinate in the aperture plane (x, y).

Now, the complex amplitude is recorded in the Fourier plane using a collimating lens L obtained by running the F.T. on Eq. (4.1) to obtain:

$$B(u, v) = \text{F.T.}\{d(x, y) \cdot P(x, y)\} \tag{4.2}$$

Making use of the F.T. properties and convolution operations we obtain

$$B(u, v) = \text{F.T.}\{d(x, y)\} \otimes \text{F.T.}\{P(x, y)\} = s(u, v) \otimes h(u, v) \tag{4.3}$$

$s(u, v) = \text{F.T.}\{d(x, y)\}$ is the speckle pattern in the case of ideal coherent imaging. In general, the speckle pattern is affected by the PSF corresponding to the aperture used.

The PSF for linear aperture computed in [6] gives the following result:

$$h(W) = 4\pi \left\{ \frac{J_1(W)}{W} + \frac{J_0(W)}{W^2} - \frac{2\sum_i J_i(W)}{W^2} \right\} \tag{4.4}$$

where $W = k\, r_0\, \rho/f$ and $\rho = (u, v)$ are the radial coordinates in the Fourier plane.
The PSF for the quadratic aperture [9] is given as follows:

$$h(W) = 4\pi \left\{ \frac{J_1(W)}{W} + \frac{J_2(W)}{W^2} \right\} \tag{4.5}$$

For a Hamming aperture with a central dark radius r_d, the PSF is represented as follows:

$$P_{\text{ham}}(\text{Obst.})(r) = 0.54 + 0.46\cos[\beta\pi(r - r_d)] \tag{4.6}$$

The PSF is computed by running the FT on Eq. (4.6) to obtain the following equation:

$$h(r) = 0.54\delta(\rho) + 0.23 \left\{ \begin{array}{c} \delta_1[\rho - \beta\lambda(f + 2r_d)] \\ +\delta_1[\rho + (f + 2r_d)] \end{array} \right\} \tag{4.7}$$

Hence, the PSF is represented by three Dirac delta functions, one at the center and the other two symmetric delta functions found at $\rho = \pm\beta\lambda(f + 2r_d)$ from the center of the diffraction pattern.

(i) For an open circular aperture, the known PSF is computed as

$$h(W) = \left\{ \frac{2J_1(W)}{W} \right\} \tag{4.8}$$

Consequently, the intensity of the speckle image is obtained by taking the modulus squared of Eq. (4.3):

$$I(u, v) = |s(u, v) \otimes h(u, v)|^2 \tag{4.9}$$

where the PSF is dependent upon the aperture and computed from Eqs. (4.4), (4.5), (4.7) and (4.8).

Now, the distribution of speckle sizes in a speckle pattern is found by examining the FWHM of the autocorrelation function of the speckle intensity pattern which is computed as follows:

$$c = < I(u, v) \cdot I^*(u, v) > \tag{4.10}$$

The contrast of the speckle images is defined as the ratio of the standard deviation σ of the intensity I to the mean intensity $<I>$ of the speckle pattern:

$$K = \frac{\sigma}{<I>} = \frac{\sqrt{<I^2> - <I>^2}}{<I>} \tag{4.11}$$

The ensemble average and standard deviation in Eq. (4.11) for the contrast are valid for a considerable number of samples or as $N \rightarrow \infty$.

Otherwise, for a selected sample of dimensions 8×8 pixels from the whole speckle image of dimensions $N \times N = 1024 \times 1024$ pixels, the ensemble average replaced by a summation ran over the sample zone, hence Eq. (4.11) becomes [15]

$$K = \frac{\sqrt{\frac{1}{(n+1)^2} \sum_{i-n/2}^{i+n/2} \sum_{j-n/2}^{j+n/2} I_{x,y}^2 - \left(\frac{1}{(n+1)^2} \sum_{i-n/2}^{i+n/2} \sum_{j-n/2}^{j+n/2} I_{x,y} \right)^2}}{\frac{1}{(n+1)^2} \sum_{i-n/2}^{i+n/2} \sum_{j-n/2}^{j+n/2} I_{x,y}} \tag{4.12}$$

4.3   Results and Discussion

Images of apertures used in computing the contrast of speckle images are plotted in Fig. 4.1a. The matrix dimensions are 1024×1024 pixels, while the radius = 256 pixels corresponds to each aperture. In the plot, circular, linear, quadratic, and obstructed Hamming apertures are shown. A magnified image of the obstructed Hamming aperture is shown in Fig. 4.1b.

The autocorrelation function is computed and plotted in Fig. 4.2a for circular and linear apertures. The FWHM = 412 pixels for the circular aperture and 258 pixels for the linear aperture. In addition, the results are plotted in Fig. 4.2b where FWHM

Fig. 4.1 a Images of apertures used in computing the contrast of speckle images **b** magnified image of the obstructed Hamming aperture

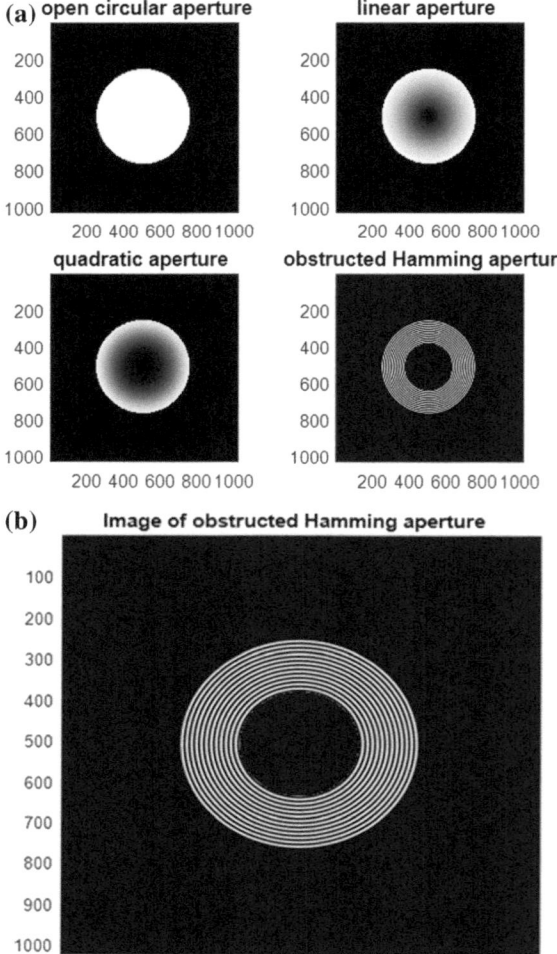

$= 182$ pixels for the quadratic aperture and FWHM $= 190$ pixels for the Hamming aperture. The results show that the bandwidth is lower for modulated apertures than for open circular apertures.

Speckle images obtained using different apertures computed via the FFT technique run upon the filled aperture with the randomly distributed function. The FT of the multiplication of the diffuser and the aperture is transformed into a convolution corresponding to the FT of each function. Hence, the speckle image is the convolution of the static speckle image and the PSF of the aperture. The speckle images for the different apertures are shown in Fig. 4.3, where the aperture radius $= 256$ pixels. The shape of the speckle images depends on the aperture distribution.

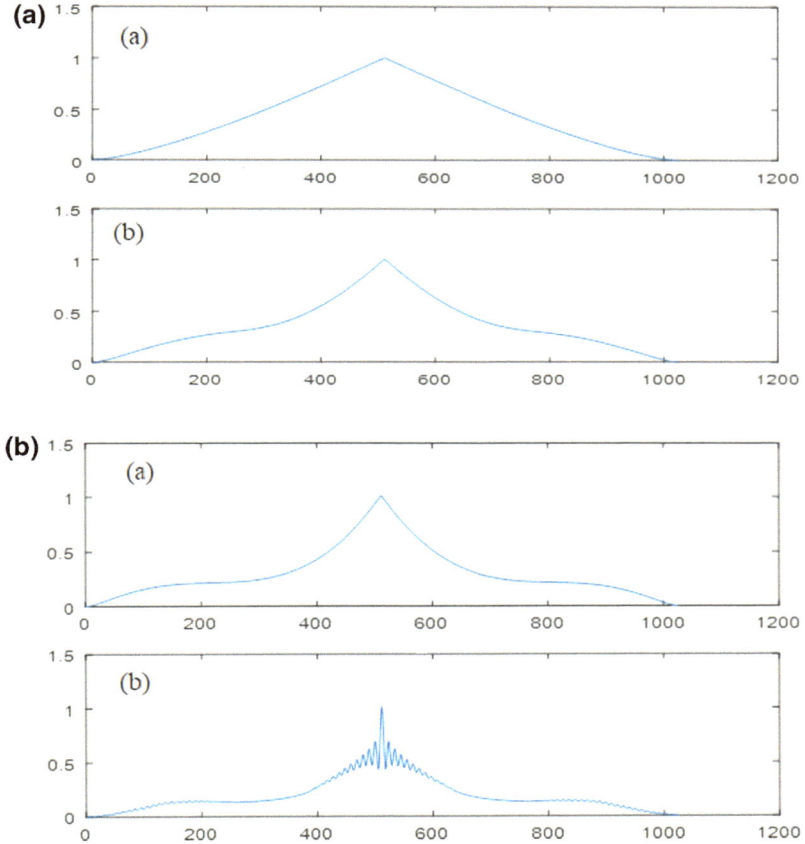

Fig. 4.2 a Autocorrelation of the circular aperture shown in (**a**) and linear aperture shown in (**b**). The FWHM $= 412$ pixels for the circular aperture while the FWHM $= 258$ pixels for the linear aperture **b** Autocorrelation of the quadratic aperture shown in (**a**) and hamming aperture shown in (**b**). The FWHM $= 182$ pixels for the quadratic aperture, while the FWHM $= 190$ pixels for the hamming aperture

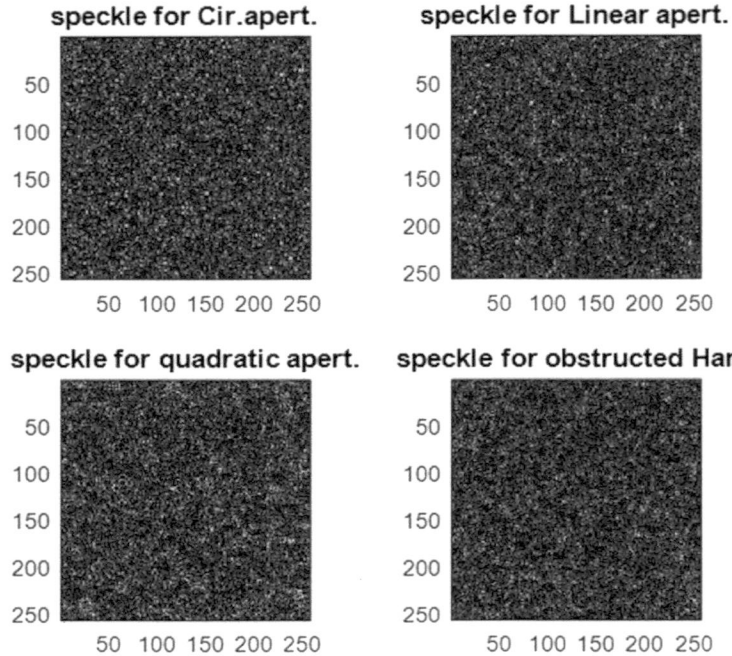

Fig. 4.3 Speckle images obtained using different apertures of the same radius = 256 pixels showing different profiles

The autocorrelation of the obtained speckle images shown in Fig. 4.3 is plotted in Fig. 4.4, where FWHM = 270, 297, 290, and 250 pixels for a → d, respectively.

The speckle image was produced from the diffuser and the quadratic aperture used in the processing, where the aperture radius = 256 pixels as shown in Fig. 4.5a. A speckle image of a sample with dimensions of 8 × 8 pixels was taken from the whole speckle image shown in Fig. 4.5b. The autocorrelation of the sampled speckle with dimensions of 8 × 8 pixels is computed and plotted in Fig. 4.5c. The FWHM results were obtained from the sampled speckle images using the different apertures shown in Fig. 4.5c–f and tabulated in Table 4.1.

For visible light and a moderate numerical aperture NA = 0.5, and the speckle size is ~ λ/NA = 1000 nm.

The speckle images were obtained using Hamming apertures of radii of 256, 128, 64, and 32 pixels which are inversely proportional to the speckle sizes for monochromatic light from the laser beam as shown in Fig. 4.6. The speckle size is $S = \lambda$/NA, where NA is proportional to the aperture radius for constant focal length f. The contrast is computed for the sampled images with dimensions of 8 × 8 pixels as a function of S. Contrast values are computed from two different samples using different apertures and fixed diffusers in the formation of the speckle image given in Table 4.2, where the aperture radius for the whole speckle images is set equal to

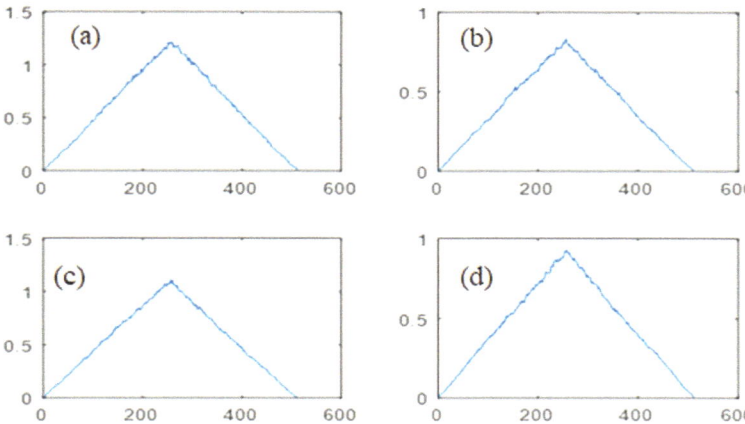

Fig. 4.4 Autocorrelation of the obtained speckle images shown in Fig. 4.3. The FWHM = was 270, 297, 290, and 250 pixels for a → d, respectively

256 pixels. In Table 4.3, the variation in contrast with respect to the aperture radius is shown for the four apertures.

Assume that the smallest speckle size = 4 pixels and the largest aperture radius = 256 pixels and that the average speckle size computed from the diffraction of the circular aperture is $S = \lambda/NA$, $\lambda f = S \times r_{max} = 4 \times 256 = 1024$ (pixels)^2. It is assumed that λ and f are constants for monochromatic illumination and a constant focal length of the objective lens. Consequently, the speckle size is inversely proportional to the aperture radius (see Table 4.4).

Four curves of contrast vs. the average speckle size are plotted using different apertures in Fig. 4.7. The plot was obtained using the inverse relation between the aperture radius and the speckle size. It is assumed that λ and f are constants that are verified for monochromatic laser illumination and for the use of certain objective lenses with variable radii. The contrast for different apertures decreased with increasing speckle size as expected.

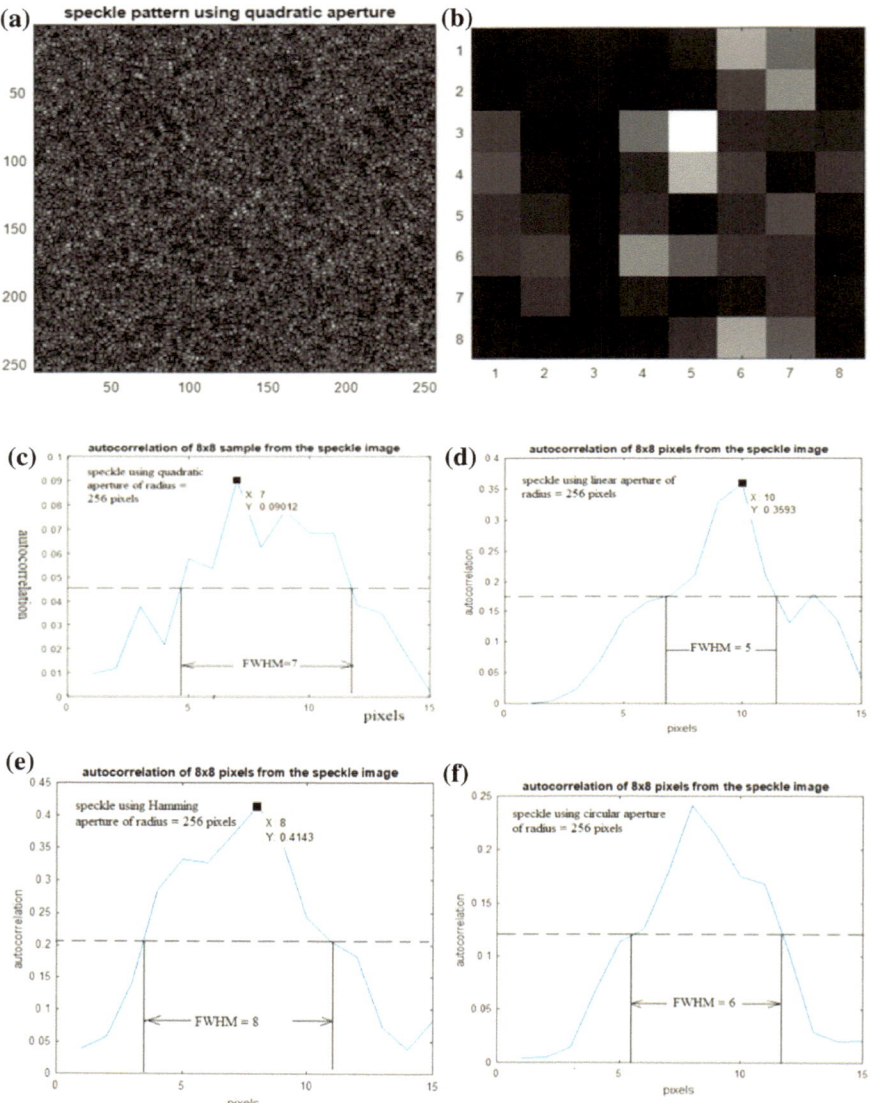

Fig. 4.5 a Speckle image produced from the diffuser and the quadratic aperture used in the processing, where the aperture radius $= 256$ pixels **b** Sampled speckle image with dimensions of 8×8 pixels taken from the whole speckle image shown in **a. c–f** autocorrelation of the sampled speckle where FWHM represents the average speckle size using the quadratic, linear, hamming, and circular apertures, respectively

Table 4.1 FWHM was obtained from the sampled images with dimensions of 8 × 8 pixels using different apertures, where the aperture radius = 256 pixels in the whole speckle image

Aperture of radius = 256 pixels	FWHM (pixels)
Linear	5
Quadratic	7
Hamming	8
Open circular aperture	6

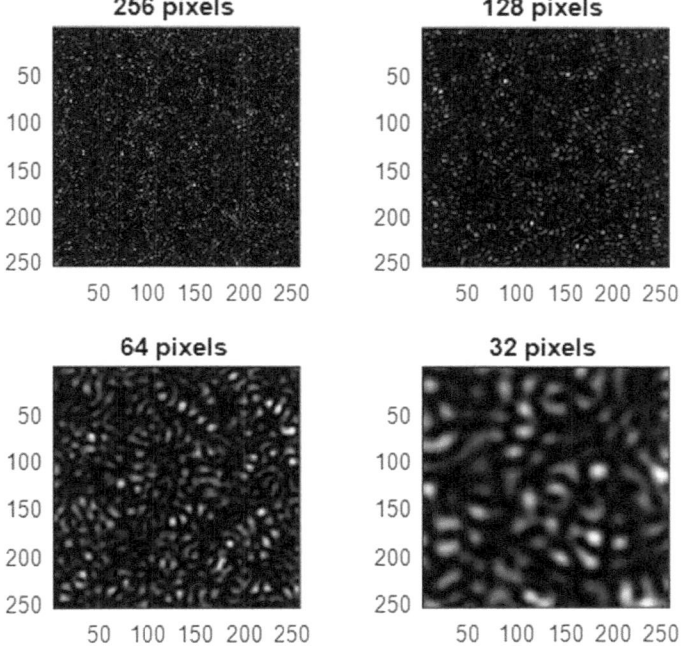

Fig. 4.6 Speckle images obtained using hamming apertures of different radii 256, 128, 64, and 32 pixels which are inversely proportional to the speckle size for monochromatic light from the laser beam. It is known that the speckle size $S = \lambda/NA$, where NA is proportional to the aperture radius for a constant focal length f. The contrast is computed for the sampled images with dimensions of 8 × 8 pixels as a function of S

Table 4.2 Contrast values were computed from two different samples using different apertures and fixed diffusers in the formation of the speckle image

Aperture	Sampled values in pixels	$<I>$	σ	$K = \sigma/<I>$
Linear	8×8	0.0967	0.0417	0.4309
	16×16	0.1054	0.0521	0.4948
Quadratic	8×8	0.0928	0.0394	0.4240
	16×16	0.079	0.0372	0.4702
Hamming	8×8	0.0646	0.0294	0.4550
	16×16	0.0727	0.0332	0.4565
Circular	8×8	0.0938	0.0374	0.3985
	16×16	0.0979	0.0447	0.4563

Table 4.3 Contrast values were computed from the 8×8 pixel sample image using different apertures and fixed diffusers in the formation of the speckle image

Aperture radius (pixels)	$K_{Circular}$	K_{Linear}	$K_{Quadratic}$	$K_{Hamming}$
256	0.4731	0.4572	0.4244	0.4671
224	0.4588	0.4176	0.4002	0.4418
192	0.4256	0.3897	0.3898	0.4173
160	0.4043	0.3743	0.3547	0.4020
128	0.3623	0.3693	0.3490	0.3780
96	0.3230	0.3194	0.3088	0.2997
64	0.3024	0.3016	0.2631	0.2022
32	0.1565	0.1578	0.2105	0.1545

Table 4.4 Relationship between the aperture radius (r) and average speckle size (S), where the multiplication of $r \times S = 1024 = $ constant

Aperture radius (pixels)	Speckle size (S) in pixels
256	4
224	4.57
192	5.33
160	6.4
128	8
96	10.67
64	16
32	32

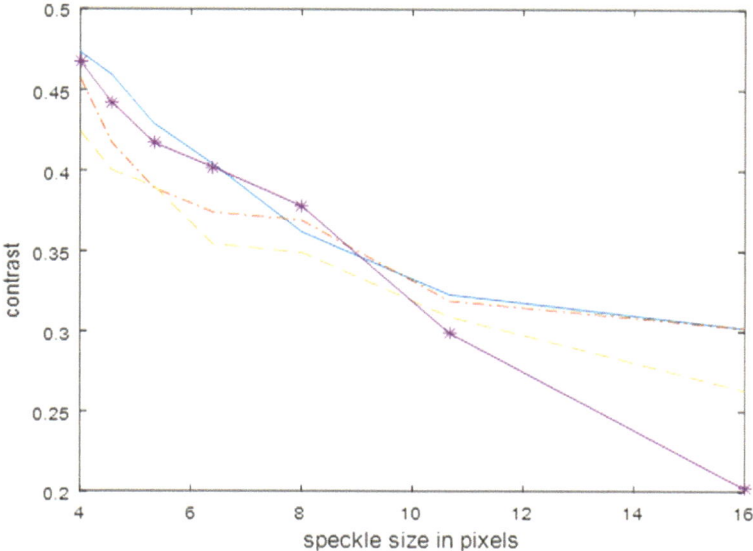

Fig. 4.7 Relation between the contrast K and the speckle size S using four different apertures. The solid line represents the circular aperture, the discontinuous dotted line (-.) stands for the linear aperture, the discontinuous line (--) represents the quadratic aperture, while the line with an asterisk (*) represents the Hamming aperture. The results were obtained using Tables 4.2 and 4.3

4.4 Conclusion

The contrast of sampled images obtained from the speckle images is correlated with the speckle size because the contrast is inversely proportional to the speckle size. In addition, the speckle size is inversely proportional to the aperture radius according to the diffraction formula obtained for the speckle pattern using a circular aperture. Hence, the aperture radius is directly proportional to the speckle contrast. Consequently, the contrast may improve when the numerical aperture increases, which result in small cutoff values in the diffraction pattern as given in Tables 4.2 and 4.3 which correspond to small speckle sizes. In addition, the resolution computed from the PSF improved by using linear or quadratic methods or a combination of both apertures and by using high NA which results in small speckle sizes.

References

1. J.W. Goodman, Statistical properties of laser speckle patterns, in *Laser Speckle and Related Phenomena*, 2nd edn, ed. by J.C. Dainty (Springer Verlag, 1984)
2. A.E. Ennos, Laser speckle, and related phenomena, in *Topics in Applied Physics*, vol. 9, ed. by J.C. Dainty (Springer Verlag, Berlin, Heidelberg, New York, 1975), p. 203

3. M. Francon, in *Laser Speckle and Applications in Optics*, 1st edn (Academic Press, 1979). ISBN: 9780323160728
4. S. Lowenthal, H. Arsenault, Image formation for coherent diffuse objects: statistical properties. J. Opt. Soc. Am. **60**, 1478–1483 (1970)
5. J.W. Goodman, in *Speckle Phenomena in Optics: Theory and Applications* (Roberts, 2006)
6. A.M. Hamed, Numerical speckle images formed by diffusers using modulated conical and linear apertures. J. Mod. Opt. **56**, 1174 (2009). https://doi.org/10.1080/09500340902985379
7. A.M. Hamed, Formation of speckle images formed for diffusers illuminated by modulated apertures (circular obstruction). J. Mod. Opt. **56**, 1633 (2009). https://doi.org/10.1080/095003 40903277792
8. A.M. Hamed, Discrimination between speckle images using diffusers modulated by some deformed apertures: simulations. Opt. Eng. **50**, 1 (2011). https://doi.org/10.1117/1.3530085
9. A.M. Hamed, Improvement of point spread function (PSF) using linear-quadratic aperture. Optik **131**, 838 (2017). https://doi.org/10.1016/j.ijleo.2016.11.201
10. A.F. Fercher, J.D. Bzuers, Flow visualization employing single-exposure speckle photography. Opt. Commun. **37**, 326–330 (1981)
11. T. Yoshimura, Statistical properties of dynamic speckles. J. Opt. Soc. Am. **3**(7), 1032–1054 (1986)
12. N. Takai, et al., Correlation distance of dynamic speckles. Appl. Opt. **22**(1), 170–177 (1983)
13. L. Allen, D.G.C. Jones, An analysis of the granularity of scattered optical maser light. Phys. Letters **7**(5), 321–323 (1963)
14. T. Asakusa, N. Takai, Dynamic laser speckles and their application to velocity measurements of the diffuse object. Appl. Phys. **25**, 179–194 (1981)
15. J. D. Briers, in *Proceedings of the Symposium on Photonics Technologies for 7th Framework Program* (Wroclaw, 12–14 October 2006)
16. M. Draijer et al., Review of laser speckle contrast techniques for visualizing tissue perfusion. Lasers Med. Sci. **24**, 639–651 (2009)
17. J.D. Briers et al., Laser speckle contrast imaging: theoretical and practical limitations. J. Biomed. Opt. **18**(6), 066018 (2013)
18. J.C. Ramirez-San-Juan et al., Impact of velocity distribution assumption on simplified laser speckle imaging equation. Opt. Express **16**(5), 3197–3203 (2008)
19. Ni Guangming, et al. Chapter 9 Optical coherence tomography in biomedicine. Biomedical Opt. Imag. **10**, 1063 (2021). https://doi.org/10.1063/9780735423794_009
20. K. Khaksari, S.J. Kirkpatrick, Combined effects of scattering and absorption on laser speckle contrast imaging. J. Biomed. Opt. **21**(7), 076002 (2016)
21. J. Liu, H. Zhang et al., Quantitative model of diffuse speckle contrast analysis for flow measurement. J. Biomed. Opt. **22**(7), 076016 (2017)
22. L. Jialin et al., Established the quantitative relationship between diffuse speckle contrast signals and absolute blood flow. Biomed. Opt. Express **9**, 4792–4806 (2018)
23. A.M. Hamed, Image processing of Ramses II statue using speckle photography modulated by a new Hamming linear aperture. Pramana—J. Phys. **94**, 120 (2020)
24. A.M. Hamed A M, Contrast of laser speckle images using some modulated apertures. Pramana—J. Phys. **95**, 122 (2021). https://doi.org/10.1007/s12043-021-02151-8

Chapter 5
Speckle Images Modulated by a New Hamming Linear Aperture

In this chapter, a new concentric Hamming linear aperture is suggested. The point spread function was computed in the case of a concentric Hamming linear aperture and compared with that corresponding to circular, conventional Hamming, and obstructed Hamming apertures. In addition, the autocorrelation corresponding to the aperture under consideration is computed and plotted. The Hamming linear aperture forms the modulated speckle images of Ramses II images using a definite random diffuser.

A diffuser limited by a circular uniform aperture is the basic element in the formation of ordinary speckle images found in the Fourier plane. Moreover, the Fourier spectrum of images from Ramses II affects the ordinary speckle forming modulated speckle images. Discrimination between the coded images in the modulated speckle images was reached by comparing the profile corresponding to each speckle image. Finally, the speckle contours and correlation of the modulated speckle images compared with the ordinary speckle images were investigated. All computations and formations of images and plots are based on fast Fourier transform (FFT) techniques using MATLAB codes.

5.1 Introduction

The speckle images are formed from the process of taking the FFT for the multiplication of the diffuser and are limited by circular or any other modulated aperture [1–10]. Hence, the speckle images are computed from the convolution product of the Fourier spectrum of the diffuser and the PSF corresponding to the aperture. The numerical Fourier holograms originated from the process of running the FFT upon the multiplication of the object coded and the diffuser [11–20]. Hence, we obtain the Fourier spectrum corresponding to the object convoluted with the Fourier spectrum of the randomly distributed function. Consequently, the common element in the

A. Hamed, *Speckle Imaging Using Aperture Modulation*,
SpringerBriefs in Applied Sciences and Technology,
https://doi.org/10.1007/978-3-031-58300-1_5

formation of both hologram and speckle images is the randomly distributed function called the diffuser which is considered an object or carrier wave in holography.

In this chapter, we investigate the images of Ramses II using speckle photography originating from the scattering of the coherent laser beam when incident upon the diffuser. We consider a new Hamming linear aperture [21] in the formation of speckle images and compare them with the corresponding speckles in the case of conventional and obstructed Hamming apertures. Speckle profiles were investigated for all the images of Ramses II. Finally, the results and discussion are given followed by the conclusion.

5.2 Theoretical Analysis

5.2.1 Computation of the PSF for a Hamming Linear Aperture

The concentric Hamming linear aperture is assumed to have a linearly varied distribution in the center surrounded by the Hamming zone. In this case, the two zones are independent, and the aperture is represented as the sum of the two zones as follows:

$$P_{H-L}(x, y) = P_L(x, y) + P_H(x, y)$$
$$= r_L + 0.54 + 0.46 \cos\left[\beta\pi\left(r - \frac{r_0}{4}\right)\right] \tag{5.1}$$

$$P_L(r) = r_L \quad \text{for} \quad \left|\frac{r_L}{r_0}\right| \leq 1/4; \ r = (x, y)$$

and

$$P_H(r) = 0.54 + 0.46 \cos\left[\beta\pi\left(r - \frac{r_0}{4}\right)\right] \quad \text{for} \ \frac{\pi}{4} \leq r \leq r_0$$

β is a parameter that has a value between zero and one. Now, the PSF is computed by applying the Fourier transform to Eq. (5.1), and we obtain the following result [20]:

$$h_{H-L}(\rho) = \text{F.T.}\{P_{H-L}(r)\} = \text{F.T.}\{r_L\} + \text{F.T.}\left\{0.54 + 0.46 \cos\left[\beta\pi\left(r - \frac{r_0}{4}\right)\right]\right\}$$

$$\text{PSF}(\rho) = \left\{\frac{J_1(\alpha\rho)}{\alpha\rho} + \frac{J_0(\alpha\rho)}{(\alpha\rho)^2} - 2\sum_i \frac{J_i(\alpha\rho)}{(\alpha\rho)^3}\right\} + \{0.54 * \delta(\rho)\}$$

$$+ 0.46 * 1/2 \left[\delta_1 \left(\rho - \frac{\beta \lambda f}{2} - \frac{\beta \lambda r_0}{4} \right) + \delta_2 \left(\rho + \frac{\beta \lambda f}{2} + \frac{\beta \lambda r_0}{4} \right) \right] \Big\}$$

$$(5.2)$$

where $\alpha = 2\pi r_0 / \lambda f$ and f is the focal length of the F.T. lens.

5.2.2 *Formation of Modulated Speckle Images*

We assume that a coherent laser beam of uniform illumination is incident upon the diffuser $d(x, y)$ limited by a circular aperture $P(x, y)$ of radius $= r_0$. In addition, an image $g(x, y)$ is placed in the same plane of the aperture. In this case, the complex amplitude is represented by the multiplication of the three functions as follows:

$$A(x, y) = d(x, y) \cdot g(x, y) \cdot P(x, y) \tag{5.3}$$

where the randomly distributed function is the diffuser represented in matrix form of dimensions $M \times N$ as follows:

$$d(x, y) = \sum_{m=1}^{M} \sum_{n=1}^{N} \exp[2\pi i (m \Delta x, n \Delta y)] \tag{5.4}$$

where $M = N$ for the square matrix.

For a circular uniform aperture,

$$\begin{aligned} P_{\text{cir}}(r) &= 1 \quad \text{for} \quad |r/r_0| \leq 1 \\ &= 0 \quad \text{for} \quad |r/r_0| > 1 \end{aligned} \tag{5.5}$$

For a conventional Hamming aperture,

$$P_{\text{ham}}(r) = 0.54 + 0.46 \cos(\beta \pi r), \quad \text{where } 0 \leq r \leq r_0 \tag{5.6}$$

Here $r = \sqrt{x^2 + y^2}$ is the radial coordinate for the above aperture.

For an obstructed Hamming aperture,

$$P(\text{obst.})_{\text{ham}}(r) = 0.54 + 0.46 \cos[(\beta \pi (r - r_d)], \quad \text{where } r_d \leq r \leq r_0 \tag{5.7}$$

r_d is the central dark radius.

The Hamming linear aperture is represented in Eq. (5.1) as

$$P_{H-L}(x, y) = r_L + 0.54 + 0.46 \cos \left[\beta \pi \left(r - \frac{r_0}{4} \right) \right] \tag{5.8}$$

Consequently, the modulated speckle formed in the focal plane of a converging lens L is obtained by operating the Fourier transform on Eq. (5.3), making use of Eqs. (5.4) and (5.5) for a circular aperture or Eqs. (5.4) and (5.6) for a conventional hamming aperture, Eqs. (5.4) and (5.7) for an obstructed hamming aperture, and finally Eqs. (5.4) and (5.8) for a Hamming linear aperture. In this case, the simple product in Eq. (5.3) is transformed into a convolution product corresponding to the Fourier spectrum of each function written in symbolic form as follows:

$$\widetilde{A}(u, v) = \widetilde{d}(u, v) \otimes \widetilde{g}(u, v) \otimes h(u, v) \tag{5.9}$$

where the Fourier spectrum corresponding to each function is written as follows:

$$\widetilde{d}(u, v) = \text{F.T.}\{d(x, y)\} \tag{5.10}$$

$$\widetilde{g}(u, v) = \text{F.T.}\{g(x, y)\} \tag{5.11}$$

$$h_1(u, v) = \text{F.T.}\{P_{\text{cir}}(x, y)\}, \quad \text{for circular aperture} \tag{5.12}$$

$$h_2(u, v) = \text{F.T.}\{P_{\text{ham}}(x, y)\}, \quad \text{for Hamming aperture} \tag{5.13}$$

$$h_3(u, v) = \text{F.T.}\{P_{\text{Obst.ham}}(x, y)\}, \quad \text{for obstructed Hamming aperture} \tag{5.14}$$

$$h_4(u, v) = \text{F.T.}\{P_{\text{ham linear}}(x, y)\}, \quad \text{for Hamming linear aperture.} \tag{5.15}$$

$\rho = \sqrt{u^2 + v^2}$ is the radial coordinate in the Fourier plane.

The ordinary speckle images obtained from the Fraunhofer diffraction of the diffuser are limited by the uniform circular aperture.

5.3   Results and Discussion

The images of the apertures under consideration are compared with the circular aperture images plotted in Fig. 5.1a for a conventional Hamming aperture, Fig. 5.1b for an obstructed Hamming aperture with an obstruction width = 64 pixels, and Fig. 5.1 for a Hamming linear aperture with an equal ratio of Hamming and linear zones. The plots corresponding to the described apertures are shown in the Fig. 5.2a–c. The total diameter = 128 pixels for all the apertures. The choice of Hamming linear aperture is considered better than the obstructed Hamming aperture since we gain more intensity from the central linear zone. In addition, both the obstructed Hamming and Hamming linear apertures yield comparable resolutions better than those obtained for a circular aperture.

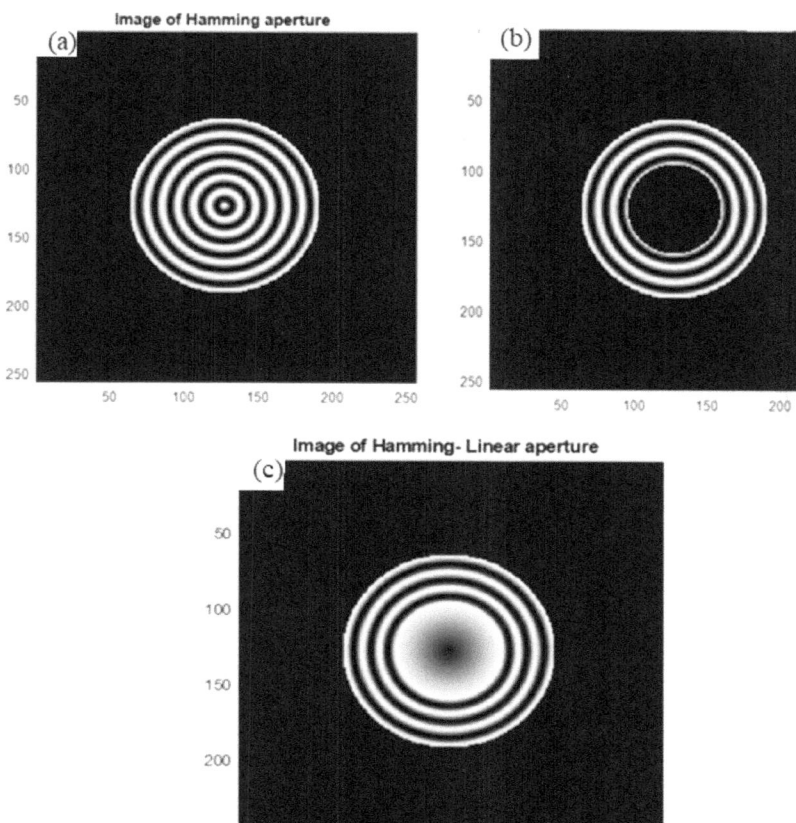

Fig. 5.1 Image of the Hamming aperture without obstruction in (**a**), hamming aperture with a central dark width $= 64$ pixels in (**b**), image of the Hamming linear aperture shown in (**c**). The diameter $= 128$ pixels in all apertures

The PSF or the Fourier spectrum corresponds to each aperture computed from the FFT technique and is plotted in Fig. 5.3a–c. The autocorrelation corresponding to each aperture is shown in Fig. 5.4a–c. Figure 5.3a shows that the PSF corresponding to the Hamming aperture is composed of a central peak surrounded by two side peaks found at $\pm$ 25 pixels as expected for the Hamming aperture. In the Fig. 5.3b, c correspond to both the obstructed Hamming and Hamming linear apertures, and the PSF or the normalized spectrum has a similar pattern to that obtained in the case of a conventional Hamming aperture but is affected by noise due to either the central dark obstruction or the linear central zone. In addition, in both cases, splitting of the peaks occurs.

The autocorrelation of the conventional Hamming aperture has a width $= 256$ pixels which is two times the aperture width as expected as shown in the Fig. 5.4a.

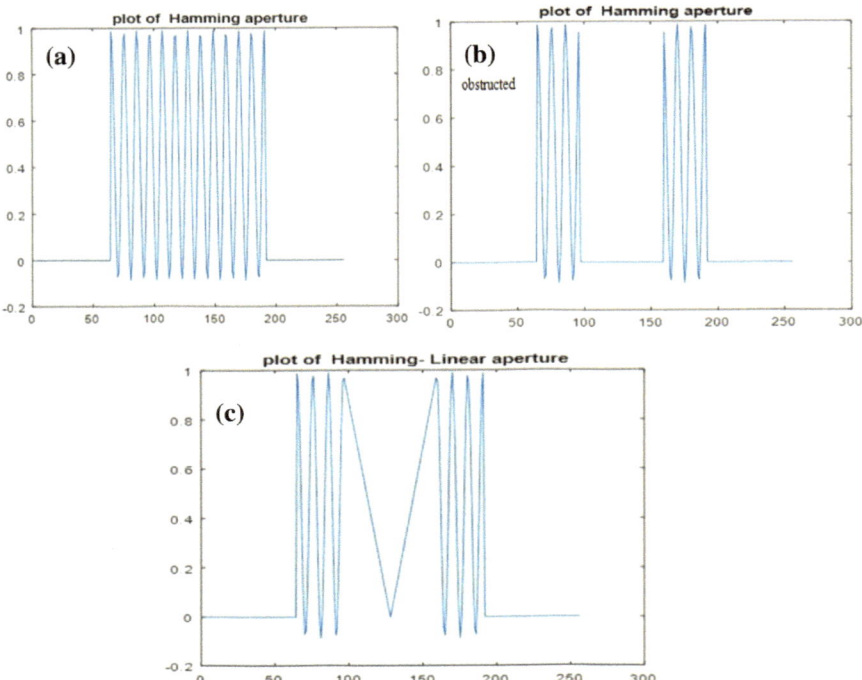

Fig. 5.2 Plot of the Hamming aperture without obstructions (**a**) and with obstructions (**b**) and hamming linear distributions in (**c**)

The autocorrelation of the obstructed Hamming aperture of the central dark zone is shown in Fig. 5.4b, and the discontinuity is shown to be proportional to the obstruction dark zone. The autocorrelation of the Hamming linear aperture as shown in Fig. 5.4c decreases harmonically linearly in proportionality with the linear central zone and again decreases harmonically until it reaches zero. The linear part in the Fig. 5.4c extends from $r = \pm 32$ pixels to $r = \pm 66$ pixels.

The modulated speckle images were obtained using the following Eqs. (5.9) and (5.15). FFT was applied to the two selected images of Ramses II. The circular, Hamming, obstructed Hamming, and Hamming linear apertures used in the formation of speckle images were created using certain diffusers illuminated with coherent laser beams. The different speckle images obtained from image 1 are shown in Fig. 5.5, while another speckle image obtained from image 2 is shown in Fig. 5.6.

The contours corresponding to the different speckle images, either ordinary or modulated, are shown in Fig. 5.7 using image 1, and in Fig. 5.8 using image 2. Figure 5.7a shows the contour of the ordinary speckle using only a diffuser and circular aperture. The other Fig. 5.7b–d corresponds to the contours for conventional Hamming, obstructed Hamming, and Hamming linear apertures, respectively, in the

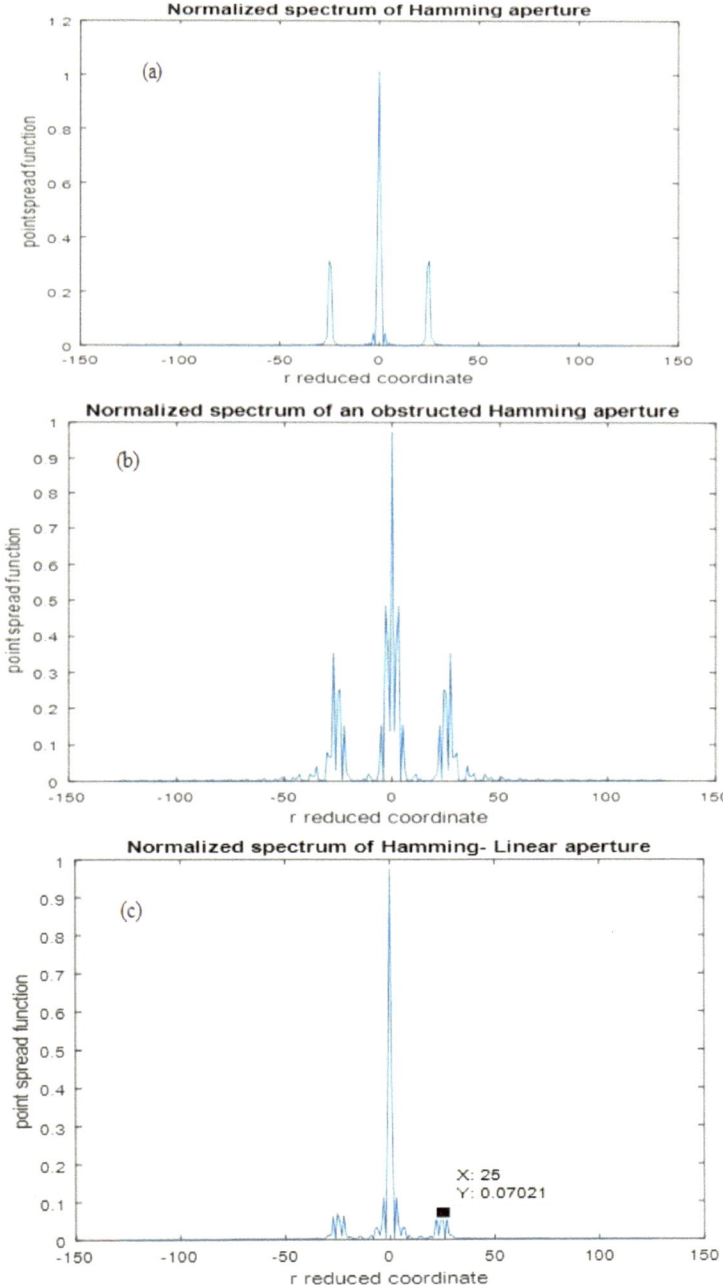

Fig. 5.3 PSF corresponds to the apertures shown in Fig. 5.1

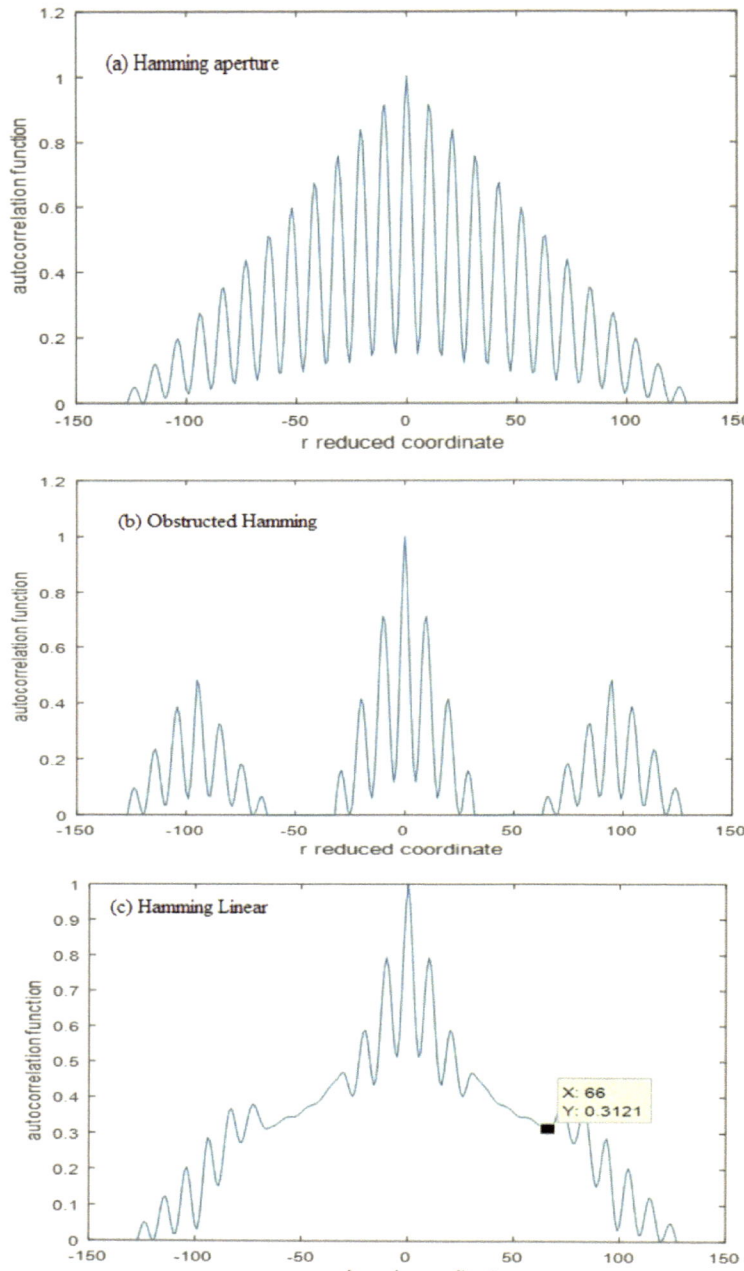

Fig. 5.4 Autocorrelation corresponding to the conventional hamming window (**a**), obstructed hamming window (**b**), and linear hamming window (**c**) is shown in Fig. 5.2

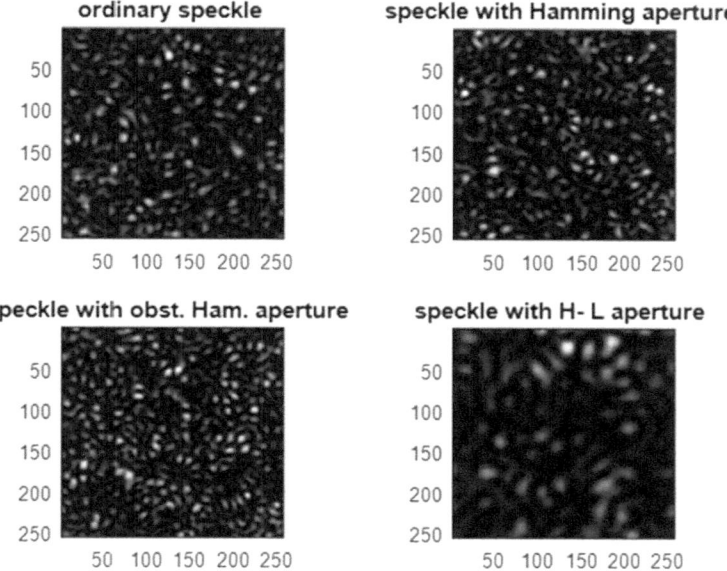

Fig. 5.5 Different speckle images formed using definite randomly distributed functions. The different apertures are shown in the speckle images using the image shown in Fig. 5.9a

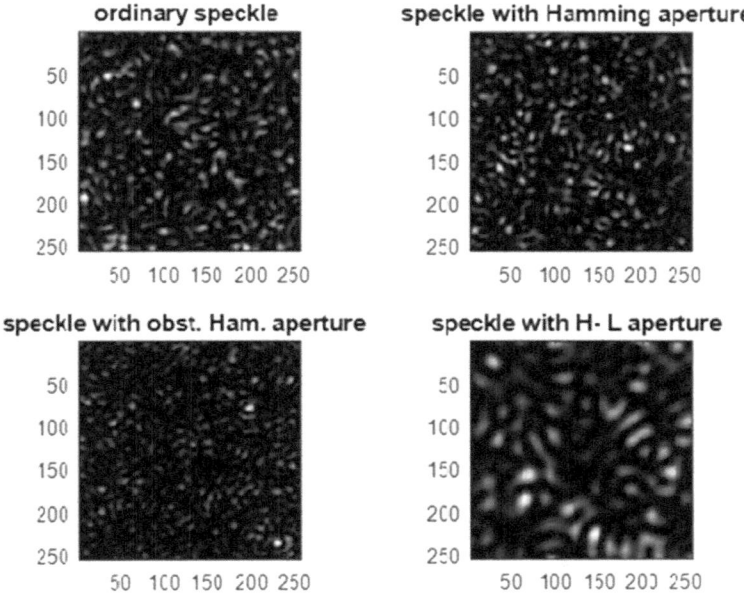

Fig. 5.6 Different speckle images formed using definite randomly distributed functions. The different apertures are shown in the speckle images using the image shown in Fig. 5.9b

presence of image 1. Similarly, modulated apertures are used to construct image 2 as shown in the Fig. 5.8b–d.

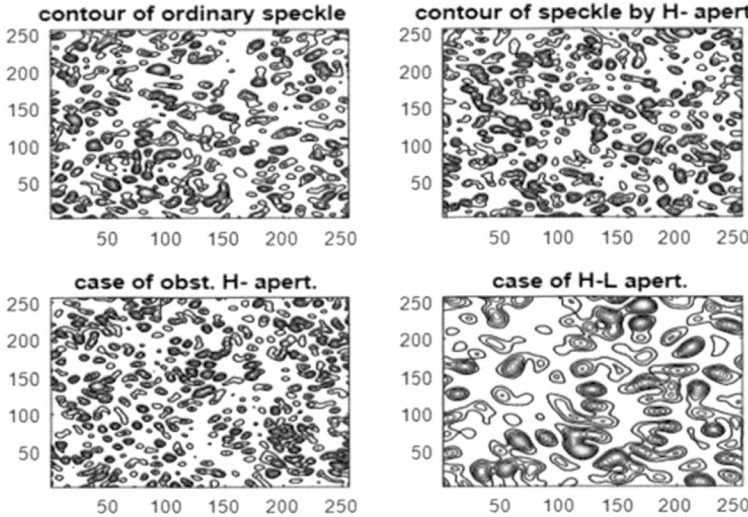

Fig. 5.7 Contours corresponding to ordinary and differently modulated speckle images using image 1 shown in Fig. 5.9a

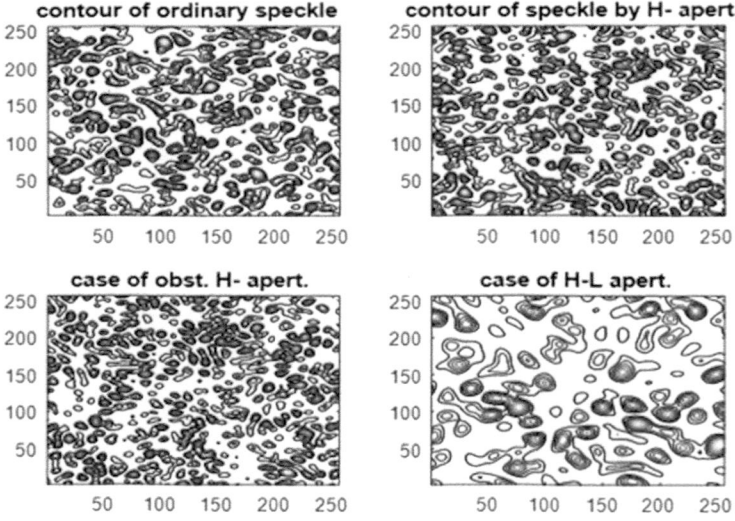

Fig. 5.8 Contours corresponding to ordinary and differently modulated speckle images using image 1 shown in Fig. 5.9b

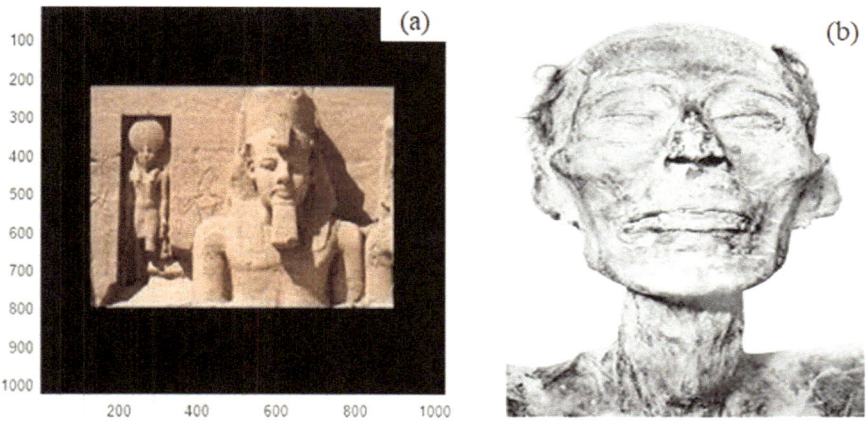

Fig. 5.9 In the L.H.S., image 1 of Ramses II is shown in (**a**) while in the R.H.S. image 2 is shown in (**b**), each rescaled to dimensions 1024 × 1024 pixels

Agglomeration of darker intensities is shown for the ordinary speckle, while more agglomeration of transparent speckle patches is shown in the presence of image 1 using the Hamming linear aperture, as shown in Fig. 5.7. Intermediate agglomeration is observed for conventional and obstructed Hamming apertures. This difference may be attributed to the diffuse scattering from the image under investigation. Hence, based on the agglomeration concept of speckle patches we can differentiate between an ordinary speckle image in the absence of an image and a modulated speckle image in the presence of an image. Another example applied to another image to justify the concept of agglomeration is shown in Fig. 5.8 which was applied to image 2. Images 1 and 2 used in the formation of speckle images and their investigation are given in Fig. 5.9a, b.

5.4 Conclusion

First, the PSF and the autocorrelation of the different Hamming apertures are either obstructed or not obstructed or Hamming linear computed and plotted.

Second, speckle images obtained from the FFT of the multiplication of the diffuser and from the Ramses II images were discriminated, and the results were limited using circular or Hamming linear apertures.

There were great differences between all the modulated speckle images. The speckle profiles corresponding to the two selected images for certain diffusers are different either in the case of a circular aperture or in the case of a Hamming linear aperture. In addition, a comparison of the profiles corresponding to the modulated speckles with the ordinary speckle profiles formed from only a diffuser and limited by a circular aperture showed a marked difference. Hence, we can confirm the presence of objects (Ramses II images) or not refer to the above method considering a reference ordinary speckle image compared with the different modulated speckle images. In addition, the autocorrelation and cross-correlations corresponding to the different modulated speckle images were investigated.

Finally, speckle contours corresponding to the modulated speckle images compared with the ordinary speckle contours are considered good discrimination methods between the different speckle images.

References

1. J.W. Goodman, in *Speckle Phenomena in Optics, Theory, and Applications* (Ben Roberts & Company, 2007)
2. L. Haibo, Speckle mechanism in holographic optical coherence imaging. Ph.D. Thesis, chapter 5, pp. 99, University of Missouri (2009)
3. A.M. Hamed, Numerical speckle images formed by diffusers using modulated conical and linear apertures. J. Mod. Opt. **56**, 1174–1181 (2009)
4. A.M. Hamed, Formation of speckle images formed for diffusers illuminated by modulated apertures (circular obstruction). J. Mod. Opt. **56**, 1633–1642 (2009)
5. E.T. Lencina, N. Bolognini, Cluster speckle structures through multiple apertures forming a closed curve. Opt. Comm. **283**, 12851290 (2010)
6. J.D. Gaskill, *Linear Systems, Fourier Transform, and Optics* (Wiley, New York, 1978)
7. A.M. Hamed et al., Analysis of speckle images to assess surface roughness. Opt. and Laser technol. **36**, 249–253 (2004)
8. A.M. Hamed, Discrimination between speckle images using diffusers modulated by some deformed apertures. Opt. Eng. **50**, 1–7 (2011)
9. A.M. Hamed, T. Al-Saeed, Image analysis of modified Hamming aperture: application on confocal microscopy and holography. J. Modern Opt. **62**, 801–810 (2015)
10. M. Francon, *Laser Speckle and Applications in Optics* (Academic Press, New York, 1979)
11. J.W. Goodman, *Introduction to Fourier optics and holography*, 3rd edn. (Roberts & Company Publishers, Greenwood Village, United States, 2005)
12. A.M. Hamed, Polychromatic image processing using thick holographic multiplexed filter. opt. Appl. **X111**, 205–213 (1983)
13. R. N. Bracewell, in *Fourier Transform and Its Applications* (McGraw-Hill, New York, 1978), Chap. 18
14. S. Trester, Computer simulated holography and computer-generated holograms. Am. J. Phys. **64**, 472–476 (1996)
15. A.M. Hamed, Scanning holography using a modulated linear pupil: simulation. Optics Photonics J. **1**, 52–58 (2011)
16. A.M. Hamed, Compromising of resolution and contrast using quadratic aperture in scanning holographic imaging. Int. J. Phot. Opt. Tech. (IJPOT) **2**, 18–23 (2016)
17. A.M. Hamed, M. Saudy, Holographic imaging of Argon plasma images. Opt. Phot. J. **4**, 136–142 (2014). https://doi.org/10.4236/opj.2014.46014
18. A.M. Hamed, Investigation of SIDA virus (HIV) images using interferometry and speckle techniques. Int. J. Innovative Res. in Comp. Sci. and Tech., **4**, 38–45 (2016)
19. A.M. Hamed, in *Topics on Optical and Digital Image Processing using Holography and Speckle Techniques* (Publisher www. Lulu.com, 2015). ISBN: 9781329328464

20. A.M. Hamed, in *The Point Spread Function of Modulated Apertures (Application on Speckle and Interferometry Images)*. (www.lap.com, Lambert Academic Publishing, 2017). ISBN:9786202070706
21. A.M. Hamed, Image processing of Ramses II statue using speckle photography modulated by a new Hamming- Linear aperture. J. Phys. PRAMANA **94**, 126 (2020)

Chapter 6
Discrimination Between Microscopy Images Using Digital Speckle Images

In this chapter, digital speckle imaging using random diffusers is considered useful for coding microscopy images. Three images of the SIDA virus, bone marrow, and red blood cells (erythrocytes) were investigated. We apply the FFT to multiply the image and the diffuser function to convolve the Fourier spectrum of the image and the Fourier spectrum of the diffuser. This resulted in a modulated speckle pattern compared with the ordinary speckle pattern formed from the diffraction of the random diffuser alone. The contrasts for the different speckle images and their profiles are plotted for the three microscopy images. In addition, reconstruction or decoding of the images from the speckle images is performed via the inverse FT operation. The transformation of either color or grayscale images into binary images allows us to obtain better images in the reconstruction process. The reconstructed images are subjected to a rotation of 180^0 to obtain a replica of the input images. All the images used in the processing were constructed using MATLAB code.

6.1 Introduction

Surface roughness plays an important role in determining the quality of machined metal surfaces in the current engineering industry. It is defined by the average of the center line profile Ra and the standard deviation or variance STD. Meanwhile, the surface contrast equals the average divided by the STD. The mechanical method of measuring surface roughness from the profiles is based on using a mechanical stylus instrument [1–3]. This method has the disadvantage of damaging the surface. Consequently, after the invention of the laser as a coherent source, optical methods were considered noncontact methods for determining surface parameters without touching the surface. The image pattern for parabolic rough surfaces is outlined in [4].

© The Author(s), under exclusive license to Springer Nature Switzerland AG 2024   61
A. Hamed, *Speckle Imaging Using Aperture Modulation*,
SpringerBriefs in Applied Sciences and Technology,
https://doi.org/10.1007/978-3-031-58300-1_6

A photographic encoder applied to an optical processor using speckle techniques produced from rough surfaces is used for discriminating between two objects [5]. The famous method is based on speckle formation from diffuser scattering [6–9] and the two-beam interferometric method [10–12], where a rough surface is placed in one of the two arms of the Michelson interferometer. Other methods for determining surface roughness are based on speckle correlation, speckle contrast techniques, and image processing [13–19]. Recently, image processing of objects using different modulated apertures and speckle imaging has been presented [20–34].

In this chapter, we suggest a method of object discrimination based on the speckle technique and image processing since for the same diffuser, different objects will give different contrasts. In addition, we reconstruct the object information from the speckle image using the FFT technique. The analysis is given followed by the results and discussion and finally the conclusion.

6.2   Analysis

Considering a collimated laser beam of uniform illumination, spatial filtering is used to ensure transparency which represents the input image. The rough surface used as a randomly distributed object may be considered a statistical variation in the random component of the surface height relative to a certain reference surface. Therefore, the random object used in this study is represented as follows:

$$d(x, y) = \exp\{j(2\pi)\text{rand}(x, y)\} \tag{6.1}$$

The multiplication of the object and the random diffuser limited by the circular aperture is represented as follows:

$$T(x, y) = d(x, y).A(x, y).P(x, y) \tag{6.2}$$

The aperture is represented as follows:

$$P(x, y) = 1; \text{ for } \sqrt{x^2 + y^2} = \rho \leq \rho_0 \tag{6.3}$$

where ρ_0 is the aperture radius. Operating FT on Eq. (6.2), we obtain the following result for the modulated speckle image:

$$\tilde{T}(u, v) = \text{F.T}\{T(x, y)\} = \text{F.T.}[d(x, y)] \otimes \text{F.T.}[A(x, y)] \otimes \text{F.T.}[P(x, y)]. \tag{6.4}$$

Or

$$\tilde{T}(u, v) = \tilde{d}(u, v) \otimes \tilde{A}(u, v) \otimes h(u, v) \tag{6.5}$$

$h(u, v) = \text{F.T.}[P(x, y)] = \frac{2J_1(w)}{w}$, where $w = \frac{2\pi\rho_0 r}{\lambda f}$ is the reduced coordinate and $r = \sqrt{u^2 + v^2}$ is the radial coordinate in the Fourier plane where the speckle images are formed.

In the absence of a circular aperture, Eq. (6.5) becomes the convolution product of the ordinary speckle and the Fourier spectrum of the image giving the speckle as follows:

$$\tilde{T}_{\text{object}}(u, v) = \tilde{d}(u, v) \otimes \tilde{A}(u, v) \tag{6.6}$$

In the absence of both the object and the pupil, the ordinary speckle is only the FT of the diffuser function written as follows:

$$\tilde{T}_{\text{ordinary}}(u, v) = \tilde{d}(u, v) \tag{6.7}$$

For image decoding, it is sufficient to apply the FFT to Eq. (6.5) to obtain reconstructed images from the speckle images.

6.3 Results and Discussion

The speckle pattern and its profile at the center of the speckle pattern at 128 pixels corresponding to the image of the SIDA virus, which was generated by making use of a uniform circular aperture with a radius of 128 pixels are shown in Fig. 6.1. The speckle pattern and the profile at the center of the speckle pattern at 128 pixels corresponding to the image of the bone marrow using a uniform circular aperture with a radius of 128 pixels are represented in Fig. 6.2. The profiles of the speckle images corresponding to the two different images (SIDA virus and bone marrow) are different. Different profiles of the speckle images shown in Figs. 6.1 and 6.2 at 100, 60, 40, and 20 pixels corresponding to the SIDA virus image and bone marrow image are plotted as in Fig. 6.3a, b. The four plots corresponding to the SIDA virus image are different from the four plots corresponding to the bone marrow image.

The difference among the three images was clear to the naked eye when we compared the speckle images corresponding to the different SIDA virus, bone marrow, and red blood cell images as shown in Fig. 6.4a–c. The profiles at the center of the speckle pattern recorded in the absence of the circular aperture at 128 pixels are obtained from the FFT corresponding to the images using the same diffuser. In this case, the PSF of the aperture does not affect the speckle distribution; rather only the Fourier spectrum of the image is the sole parameter that affects the distribution. The speckle patterns obtained from the FFT operation on the multiplication of the diffuser and the image are shown in Fig. 6.5a for the SIDA virus image in the absence of an aperture. Different speckle images of the bone marrow are shown in Fig. 6.5b, while the speckle image corresponding to the red cell image is shown in Fig. 6.5c.

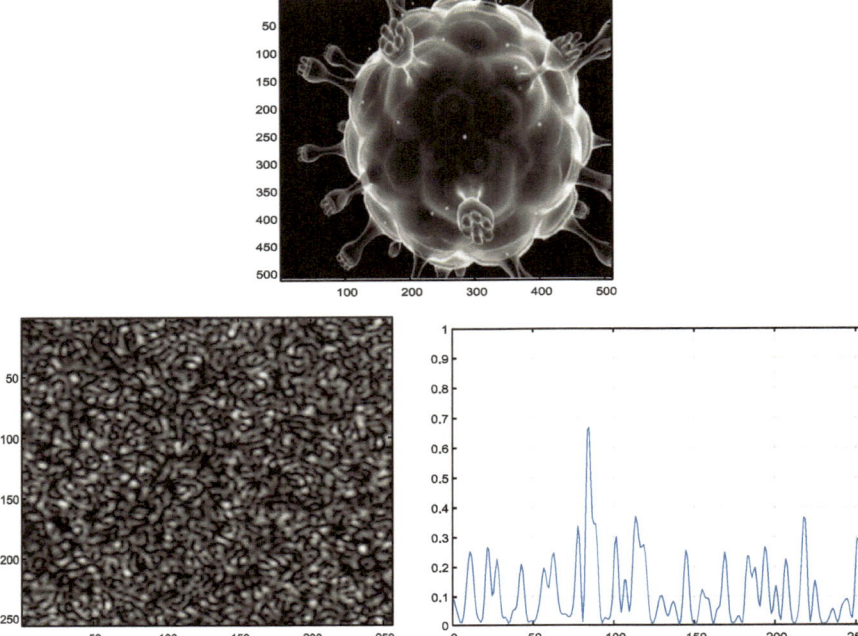

Fig. 6.1 Speckle pattern and the profile at the center of the speckle pattern at 128 pixels corresponding to the image of the SIDA virus using a uniform circular aperture with a radius of 128 pixels

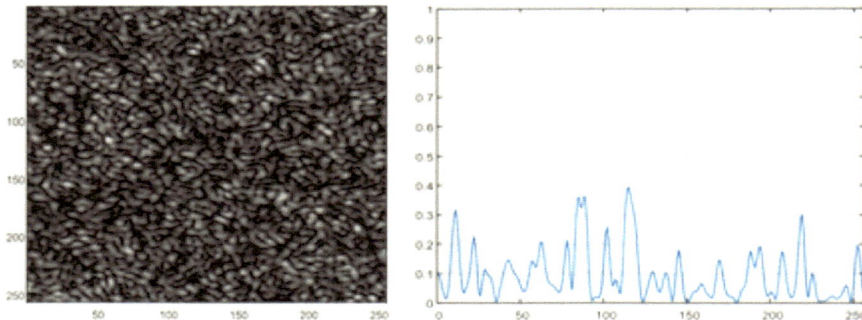

Fig. 6.2 Speckle pattern and the profile at the center of the speckle pattern at 128 pixels corresponding to the image of bone marrow using a uniform circular aperture with a radius of 128 pixels

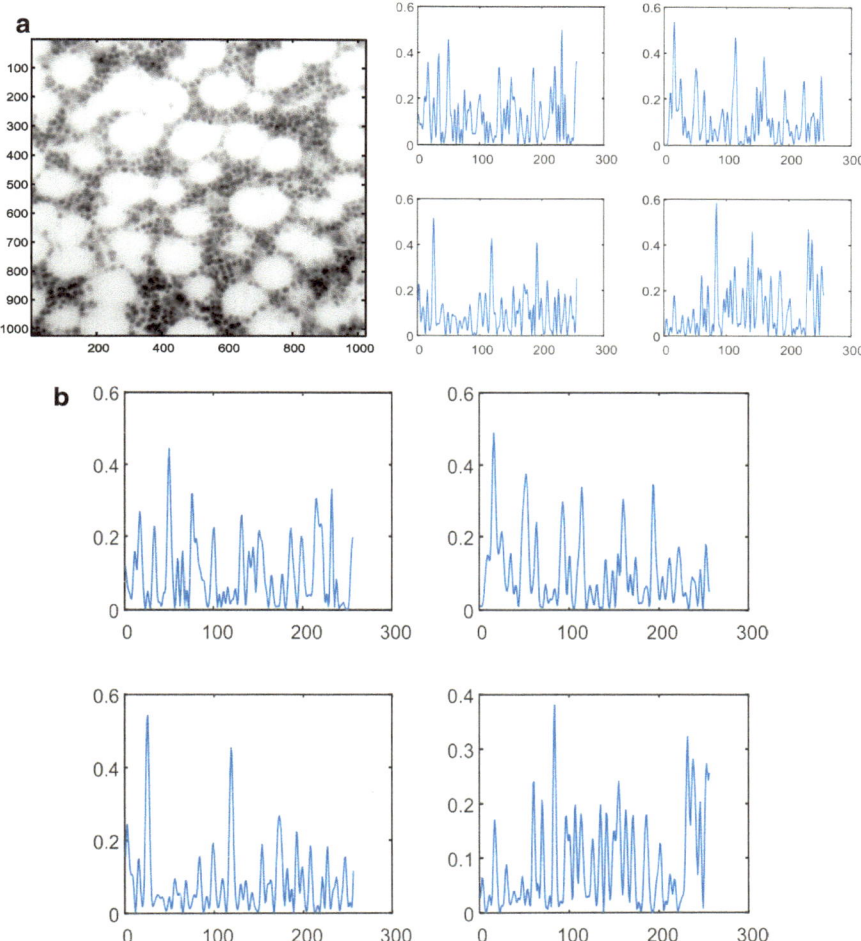

Fig. 6.3 **a** Speckle pattern profiles at 100, 60, 40, and 20 pixels corresponding to the SIDA virus image **b** speckle pattern profiles at 100, 60, 40, and 20 pixels corresponding to the bone marrow image

Image processing was applied to red blood cells and bone marrow images. In Fig. 6.6a, the original grayscale image of blood cells is shown. The reconstructed image is shown in Fig. 6.6b which is affected by noise originating from the random diffuser. The reconstruction is made from the corresponding speckle image shown in Fig. 6.5c. A trial to improve the reconstructed image is performed, and the filtered image reconstructed from the speckle pattern after eliminating the diffuser image is shown in Fig. 6.6c.

To avoid the great noise that affects the reconstructed image we transform the input grayscale or color image into a b/w image as shown in Fig. 6.6d, where the level

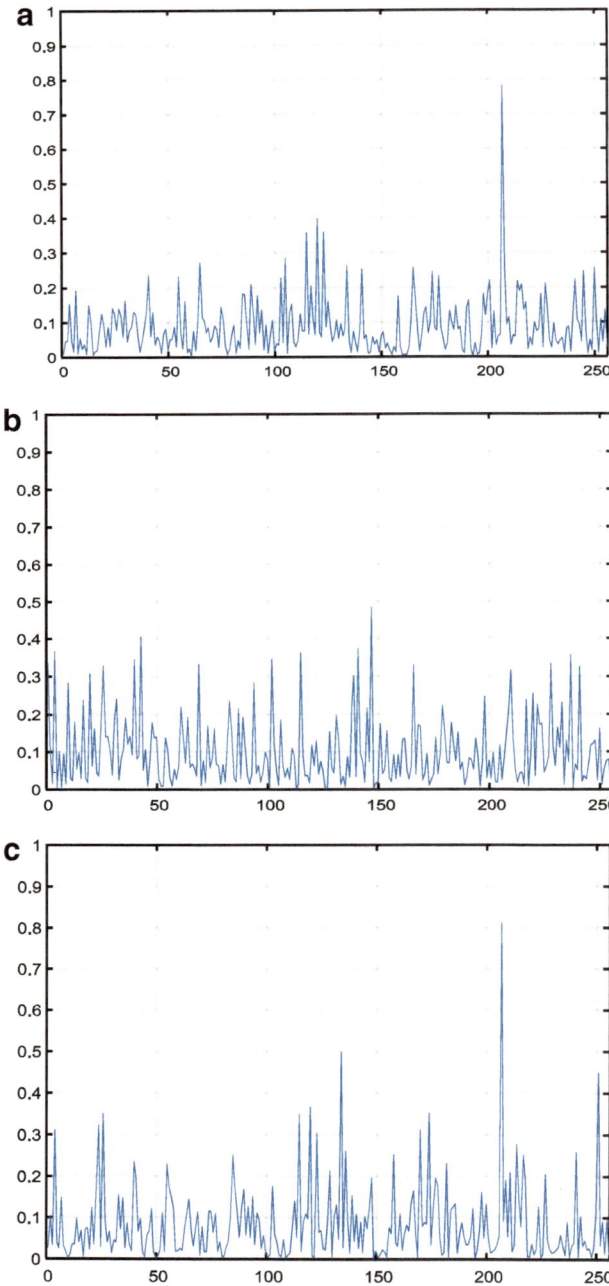

Fig. 6.4 a Profile at the center of the speckle pattern at 128 pixels corresponding to the image of the SIDA virus without an aperture **b** profile at the center of the speckle pattern at 128 pixels corresponding to the image of bone marrow without an aperture **c** profile at the center of the speckle pattern at 128 pixels corresponding to the image of blood cells without an aperture

Fig. 6.5 **a** Speckle pattern corresponding to the image of the SIDA virus using a diffuser in the absence of an aperture **b** speckle pattern corresponding to the image of bone marrow using a diffuser in the absence of an aperture **c** speckle pattern corresponding to the image of blood cells obtained using a diffuser in the absence of an aperture

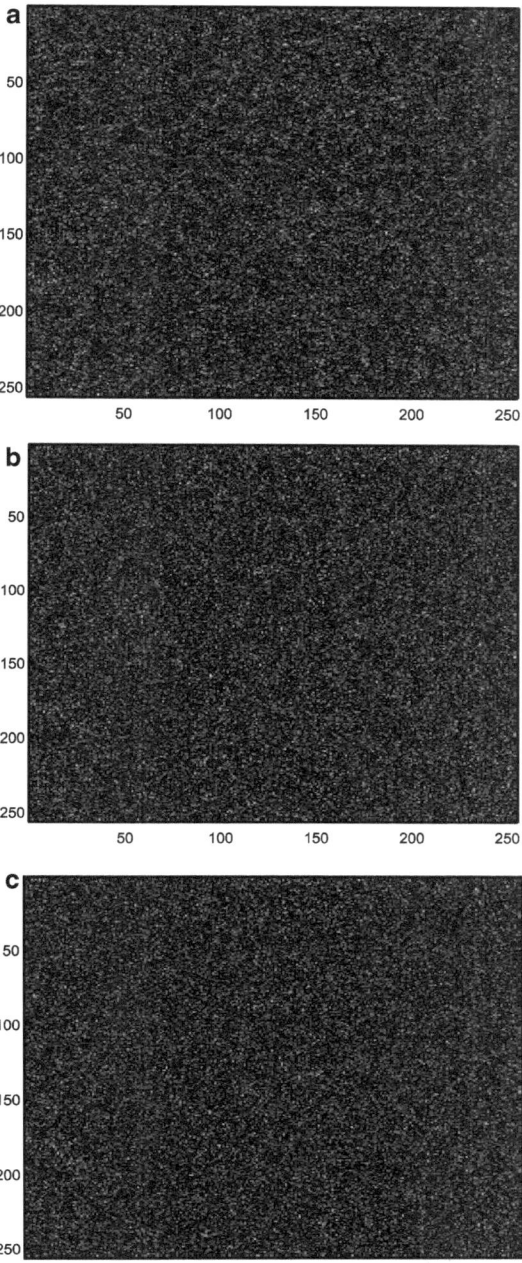

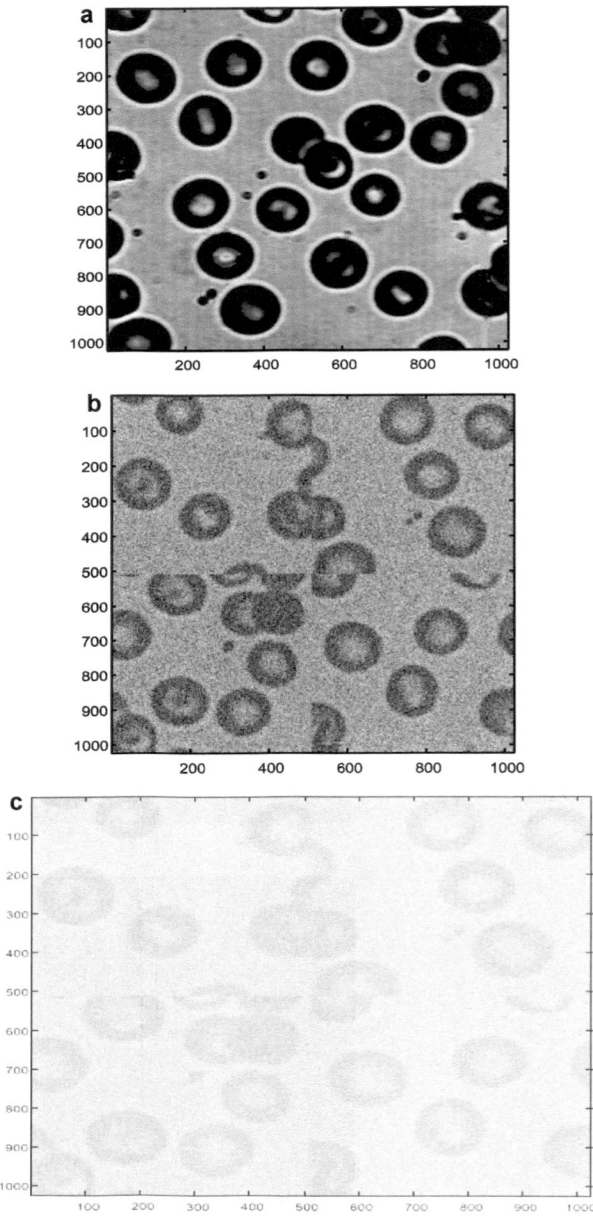

Fig. 6.6 **a** Original image of blood cells. **b** reconstructed image from the speckle pattern shown in **c**. **c** filtered image reconstructed from the speckle pattern after eliminating the diffuser image **d** binary image of blood cells corresponding to the grayscale image shown in Fig. 6.2, where the level of transformation $= 0.5$ **e** a reconstructed grayscale image of blood cells from the speckle pattern obtained by transforming the grayscale input image into a binary image as shown in Fig. 6.6d **f** four segments of the reconstructed grayscale image shown in Fig. 6.6d were combined after the rotation of each segment by $180°$. It resembles the original input image

Fig. 6.6 (continued)

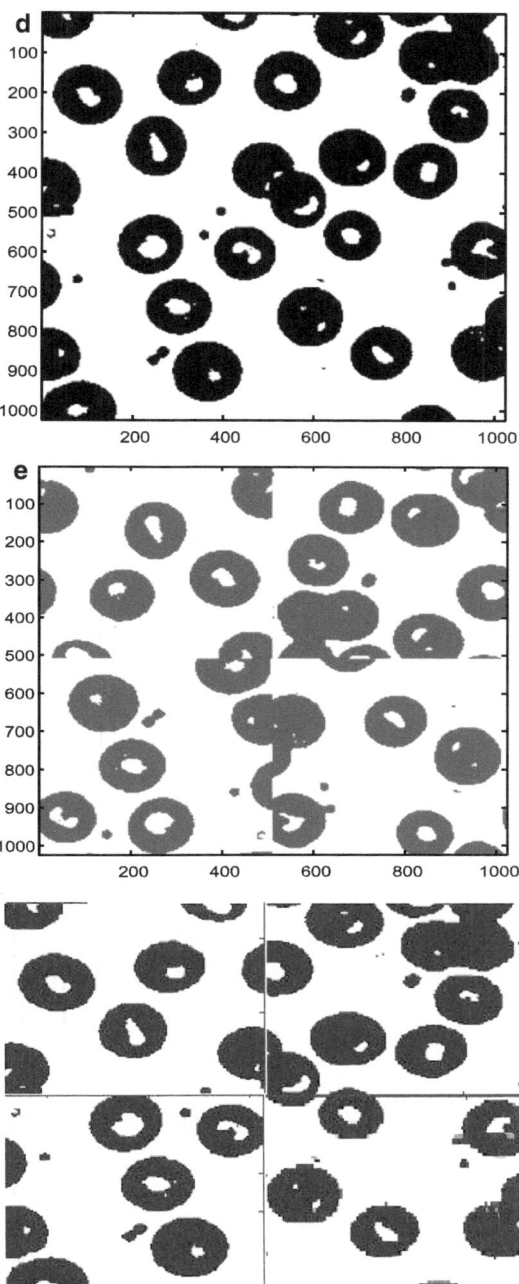

of transformation $= 0.5$. The reconstructed grayscale image of blood cells obtained from the speckle pattern after transforming the grayscale input image into a binary image is shown in Fig. 6.6e. A combined four segments of the reconstructed grayscale image shown in Fig. 6.6d after the rotation of each segment by 1800 are plotted in Fig. 6.6f. In this case, an improved similar image is obtained. Similar results are obtained for the reconstructed bone marrow image in Fig. 6.7a–c. In Table 6.1, the contrast corresponding to different speckle images is given.

These parameters are the basic factors that affect the surface profile, and the first parameter (amplitude) is defined as the surface roughness. We assume that the height profile of a given surface is a single-valued function. It has a point coordinate $h(x)$. There are two measures of surface roughness:

(i)   The average height of the profile Ra is given as follows:

$R_a = < h > = \frac{1}{L} \int_0^L h(x) \, dx$; L is the transverse measured length.

ii)   The other measuring parameter is the root mean square surface roughness or the standard deviation (STD) and is defined as

$$\sigma = R_{rms} = \sqrt{\int_0^L Y^2(x) dx},$$

where $Y(x) = h(x) - < h >$ is the deviation from the mean surface roughness. The root mean square surface roughness describes the fluctuations in surface height around an average surface height. The image contrast is computed using Ra and σ as follows: $C = Ra/\sigma$.

6.4 Conclusion

First, it is concluded that the modulated speckle patterns corresponding to different microscopy images have different profile characteristics. Second, the decoding process of the modulated speckle pattern is verified by applying the FFT technique which yields moderate reconstructed images affected by diffuser noise. Finally, the grayscale or color image is transformed into a binary image which is coded using a random diffuser to obtain the modulated speckle. This approach yields better-reconstructed images for decoding. In all the cases, the modulated speckle images are computed from the convolution product of the Fourier spectrum of the input image and the ordinary speckle pattern. The ordinary speckle pattern is obtained from the FT of the diffuser alone.

Fig. 6.7   a Binary image of a bone marrow grayscale image shown in Fig. 6.2, where the level of transformation $= 0.7$ **b** reconstructed grayscale image of bone marrow from the speckle pattern obtained by transforming the grayscale input image into a binary image. If we rotate each segment by $180°$ we obtain the exact input image as shown in Fig. 6.7c **c** reconstructed image of bone marrow after rotation of each segment by $180°$ giving a replica of the input image from the speckle pattern shown in Fig. 6.5b

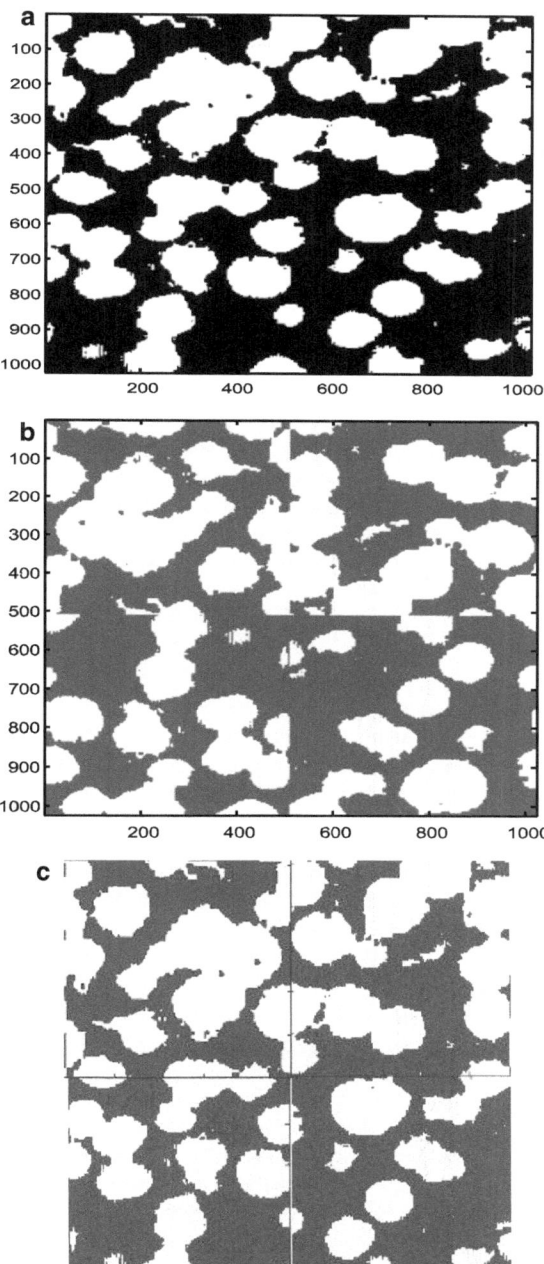

Table 6.1 Contrast corresponding to different speckle images

Image	Contrast
Image 1 (SIDA virus)	0.7462
Speckle pattern for image 1 with circular aperture	0.4508
Speckle pattern for image 1 without the aperture	0.4060
Image 2 (bone marrow)	0.3908
Speckle pattern for image 2 with circular aperture	0.4232
Speckle pattern for image 2 without the aperture	0.4263
Speckle pattern for the diffuser alone	0.4519

References

1. M.A. Sander, in *A Practical Guide to the Assessment of Surface Texture* (Germany, 1991)
2. E.S. Gadelmawla, M.M. Koura, T.M.A. Maksoud, I.M. Elewa, H.H. Soliman, Roughness parameters. J. Mater. Process. Technol. **123**, 33–145 (2002)
3. T.C. Poon, B. Bhushan, Comparison of surface roughness measurement by stylus profiler, AFM and non–contact optical profiler. Wear **190**, 77–88 (1995)
4. A.M. Hamed, M. El Shabshiry, Theoretical study of image patterns. Opt. Applicata. **13**, 317–320 (1983)
5. N. Barakat, A.M. Hamed, H. El Ghandoor et al., A photographic encoder applied to an optical processor using speckle techniques. J. Modern Opt. **38**, 203–208 (1991)
6. J. Ohtsubo, T. Asakura, Statistical properties of speckle intensity variations in the diffraction under the illumination of partially coherent light. Nouv. Rev. Opt. **6**, 189 (1975)
7. J. Ohtsubo, H. Fujii, T. Asakura, Surface roughness measurement by using speckle pattern. Jpn. J. Appl. Phys. **14**, 293 (1975)
8. H. Fujii, J. Uozumi, T. Asakura, Computer simulation study of image speckle pattern with relation to object surface profile. J Opt Soc. Amer. **66**, 1222 (1976)
9. A.M. Hamed, H. El-Ghandoor, F. El-Diasty, M. Saudy, Analysis of speckle images to assess surface roughness. Opt. Laser Technol. **36**, 249–253 (2004)
10. R. Balamurugan, S. Muruganand, Displacement measurement and study of surface roughness using laser speckle technique. Proc. of the Int. conf. on Intelligent Systems and Image. https://doi.org/10.12792/icisip2014.079
11. R.K. Erf, in *Speckle Metrology* (Academic Press, New York, 1978)
12. E. Kayahan, H. Oktem, F. Hacizade, H. Nasibov, O. Gundogdu, Measurement of surface roughness of metals using binary speckle image analysis. Tribol. Internat. **43**, 307–311 (2010)
13. U. Persson, Measurement of surface roughness on a rough machined surface using speckle correlation and image analysis. Wear **160**, 221–225 (1993)
14. B. Ruffung, Application of speckle-correlation methods to surface roughness measurement: a theoretical study. J. Opt. Soc. Am. **A3**, 1297–1304 (1986)
15. S.L. Toh, C. Quan, K.C. Woo, C.J. Tay, H.M. Shang, Whole field surface roughness measurement by laser speckle correlation technique. Opt. Laser Technol. **33**, 427–434 (2001)
16. U. Persson, Surface roughness measurement on machined surfaces using angular speckle correlation. J. Mater. Process. Technol. **180**, 233–238 (2006)
17. A.M. Hamed, M. Saudy, Computation of surface roughness using optical correlation. Pramana J. Phys. **68**, 831–842 (2007)
18. U. Persson, Real-time measurement of surface roughness on ground surface using the speckle-contrast technique. Opt Laser Eng. **17**, 61–67 (1992)
19. L.C. Leonard, V. Toal, Roughness measurement of metallic surfaces based on the laser speckle contrast method. Opt Laser Eng. **30**, 433–440 (1998)

20. A.M. Hamed, Numerical speckle images formed by diffusers using modulated conical and linear apertures. J. Modern Opt. **56**, 1174–1181 (2009)
21. A.M. Hamed, Formation of speckle images formed for diffusers illuminated by modulated apertures (circular obstruction). J. Modern Opt. **56**, 1633–1642 (2009)
22. A.M. Hamed, Discrimination between speckle images using diffusers modulated by some deformed apertures: simulations. J. Opt. Eng. **50**, 1–7 (2011)
23. A.M. Hamed, Computer generated quadratic and higher order apertures and its application on numerical speckle images. Opt. Phot. J. **1**, 43–51 (2011)
24. A.M. Hamed, Recognition of direction of new apertures from the elongated speckle images: simulation. Opt. Phot. J. **3**, 250–258 (2013)
25. A.M. Hamed, Study of the graded index and truncated apertures using speckle images. Pre. Inst. Mech. PIM **3**, 144–152 (2014)
26. A.M. Hamed, Tarek. M. Al-Saeed, Processing of mammographic images using speckle technique. Int. J. Comp. Eng. IJCER **4**, 56–62 (2014)
27. A.M. Hamed, Discrimination between normal and diseased stomach using speckle imaging. Int. J. Innovative Res. Eng. Management (IJIREM) **3**, 125–133 (2016)
28. A.M. Hamed, Investigation of SIDA Virus (HIV) images using interferometry and speckle techniques. Int. J. Innovative Res. Comp. Sci. Tech (IJIRCST) **4**, 38–45 (2016)
29. A.M. Hamed, A modified Michelson interferometer and an application on microscopic imaging. IJPOT **3** (2017)
30. A.M. Hamed, Processing of the retinal artery image using higher orders of two-beam interference. IJPOT **3** (2017)
31. A.M. Hamed, Recognition of some modulated apertures using the Cascaded Fabry-Perot Interferometer (CFPI). Int. J. Innovative Res. Eng. Management (IJIREM) **5**, 173–181 (2018)
32. A. M. Hamed, in *Polychromatic Image Processing (Laser Applications)*, 1st edn (Cairo, A.R. Egypt, 1998), p. 186. ISBN 977-19-6202-7
33. A.M. Hamed, in *Topics on Optical and Digital Image Processing Using Holography and Speckle Techniques*, 1st edn (www.lulu.com, July 6, 2015). ISBN 9781329328464
34. A.M. Hamed, The PSF of some modulated apertures, in *Application on Speckle and Interferometry Images* (Lambert Academic Publishing, 2017). www.lap.com, ISBN: 9786202070706

Chapter 7
Speckle Images Using Concentric Black and White Hexagonal Pupils

I constructed a new concentric black and white (B/W) hexagonal aperture, and we resized the image in a 1024 × 1024 pixel matrix using MATLAB code.

I derived a new formula for the point spread function (PSF) corresponding to the B/W hexagonal aperture. We computed the PSF using the fast Fourier transform (FFT) and compared the results with those for a transparent hexagonal shape. In addition, we computed the coherent transfer function (CTF) for two symmetric objectives provided with the B/W hexagonal aperture.

I showed that the full width of the central lobe is smaller than that corresponding to the transparent hexagonal aperture, hence improved resolution is attained in the case of the B/W hexagonal aperture. In addition, we observed significant peaks in the diffraction pattern corresponding to the B/W concentric hexagonal aperture. The PSF is dependent on the geometry of the B/W hexagonal aperture.

The coherent transfer function (CTF) or the autocorrelation corresponding to the B/W concentric hexagonal aperture has triangular decaying fringing compared with that corresponding to the uniform hexagonal and circular apertures.

Finally, speckle formation using modulated hexagonal apertures combined with a diffuser is presented.

7.1 Introduction

The point spread function (PSF) in optical microscopy, either conventional or confocal, is considered important in resolution computation. The PSF is defined as the Fourier transform of the aperture corresponding to the microscope objective. Early results showed that the cutoff spatial frequency in the diffraction plane is dependent on the aperture and the wavelength of light [1]. Considering the monochromatic coherent illumination of the microscope, the PSF is dependent mainly on the aperture.

© The Author(s), under exclusive license to Springer Nature Switzerland AG 2024
A. Hamed, *Speckle Imaging Using Aperture Modulation*,
SpringerBriefs in Applied Sciences and Technology,
https://doi.org/10.1007/978-3-031-58300-1_7

Hence, modulation studies based on aperture modification improve the resolution of microscope beams Hamed [2–9].

Other methods of modulation based on structured illumination microscopy (SIM) are among the most significant wide field super resolution optical imaging techniques. Conventional SIM utilizes a sinusoidal structured pattern to excite the fluorescent sample, which eventually downregulates higher spatial frequency sample information within the diffraction-limited passband of the microscopy system as outlined by Heintzman [10] and Krishnendu et al. [11]. Stochastic optical reconstruction microscopy (STORM) is a widely used super resolution technique based on the principle of single-molecule localization and was described by Jian Quan et al. [12]. STORM achieves a spatial resolution of 20–30 nm, a tenfold improvement compared with that of conventional optical microscopy.

A recent review article concerning the development of microscopy for super resolution in confocal microscopy was presented by Sheppard [13]. Recently, we provided a simple analytic solution for the PSF corresponding to the hexagonal aperture Hamed [14] using the concepts of Fourier transforms and convolution operations. The point spread function of hexagonally segmented telescopes by a new symmetrical formulation was presented in [15].

Goodman [1] and Mohamed [16] used the hexagonal aperture in the computation of the point spread function (PSF) to reduce spherical aberration. The regular hexagon is considered the sum of two trapezoids Sabatke [17] and the Fourier transform considers the division of each trapezoid into strips.

Focal modulation microscopy (FMM) has attracted significant interest in biological imaging. However, the spatial resolution and penetration depth limit the imaging quality of FMM due to the strong scattering background. Deng et al. [18] introduced the FMM with a Tai Chi aperture (TCFMM) based on diffraction theory to improve spatial resolution. They showed that the transverse resolution is improved by 61.60% and 41.37% in two orthogonal directions, and the axial resolution is improved by 29.67% compared with that of confocal microscopy (CM). These improvements in spatial resolution indicate that TCFMM has potential in deep tissue imaging.

Chen [19] and Gong [20] showed that focal modulation microscopy (FMM), which combines a spatial phase modulator with confocal microscopy, results in an improvement in spatial resolution. This technique was introduced to increase the imaging depth into tissue and rejection of background from a thick scattering object. Other apertures discussed by Wu et al. [21] obtained improvements with divided cosine-shaped apertures in confocal microscopy. Enhanced background rejection in thick tissue using focal modulation microscopy with quadrant apertures was attained by Si et al. [22].

In this chapter, I consider a new B/W concentric hexagonal aperture, obtain the PS, and discuss the resolution limit computed from the cutoff spatial frequency in the diffraction plane. In addition, I discussed the CTF in the case of this modulated hexagonal aperture. Since N represents the number of B/W zones, I select six zones only for simplicity, while the analysis is valid for any number of finite zones.

7.2 Theoretical Analysis

We summarize the PSF results for the recently presented hexagonal aperture [14] as follows:

$$P(x, y) = \text{rect}(x, y) + \text{tri}\left(x, y - \frac{2a}{3}\right) + \text{tri}\left(x, y + \frac{2a}{3}\right) \tag{7.1}$$

where $\text{rect}(x, y) = 1; \frac{x}{a} \leq 1$ and $\frac{y}{b} \leq 1$ represents a rectangle of sides a and b where $a = b$.

$$\text{tri}\left(x, y - \frac{2a}{3}\right) = 1.$$ It represents a triangle of base $2b$ and height a.

We applied the Fourier transform using Eq. (7.1), and we obtained the PSF (2023) as follows:

$$h(u, v) = \left[\frac{\sin(bu)}{bu}\right]\left[\frac{\sin(av)}{av}\right]\left\{1 + 2\left[\frac{\sin(bu)}{bu}\right]\left[\frac{\sin(av)}{av}\right]\cos\left[\left(\frac{4a}{3\lambda f}\right)v\right]\right\} \tag{7.2}$$

The intensity corresponding to the PSF is written as follows:

$$I(u, v) = |h(u, v)|^2 \tag{7.3}$$

In this chapter, we compute the PSF corresponding to six concentric zones with black and white hexagonal shapes, as shown in Fig. 7.1.

We represent the B/W hexagonal pupil as follows:

$$
\begin{aligned}
P(x, y) = &\left\{\text{rect}(x, y) + \Delta\left(x, y - \frac{2a}{3}\right) + \Delta\left(x, y + \frac{2a}{3}\right)\right\} \\
&- \left\{\text{rect}(\alpha_1 x, \beta_1 y) + \Delta\left(\alpha_1 x, y - \frac{a}{3}\right) + \Delta\left(\alpha_1 x, y + \frac{a}{3}\right)\right\} \\
&+ \left\{\text{rect}(\alpha_2 x, \beta_2 y) + \Delta\left(\alpha_2 x, y - \frac{2a}{9}\right) + \Delta\left(\alpha_2 x, y + \frac{2a}{9}\right)\right\} \\
&- \left\{\text{rect}(\alpha_3 x, \beta_3 y) + \Delta\left(\alpha_3 x, y - \frac{2a}{9}\right) + \Delta\left(\alpha_3 x, y + \frac{2a}{9}\right)\right\} \\
&+ \left\{\text{rect}(\alpha_4 x, \beta_4 y) + \Delta\left(\alpha_4 x, y - \frac{2a}{9}\right) + \Delta\left(\alpha_4 x, y + \frac{2a}{9}\right)\right\} \\
&- \left\{\text{rect}(\alpha_5 x, \beta_5 y) + \Delta\left(\alpha_5 x, y - \frac{2a}{9}\right) + \Delta\left(\alpha_5 x, y + \frac{2a}{9}\right)\right\} \tag{7.4}
\end{aligned}
$$

where $\alpha_1 = \frac{a}{2}, \alpha_2 = \frac{a}{3}, \alpha_3 = \frac{a}{4}, \alpha_4 = \frac{a}{5}, \alpha_5 = \frac{a}{6}$ and $\alpha = \beta$. For any definite number N of hexagonal zones, we write the aperture as follows:

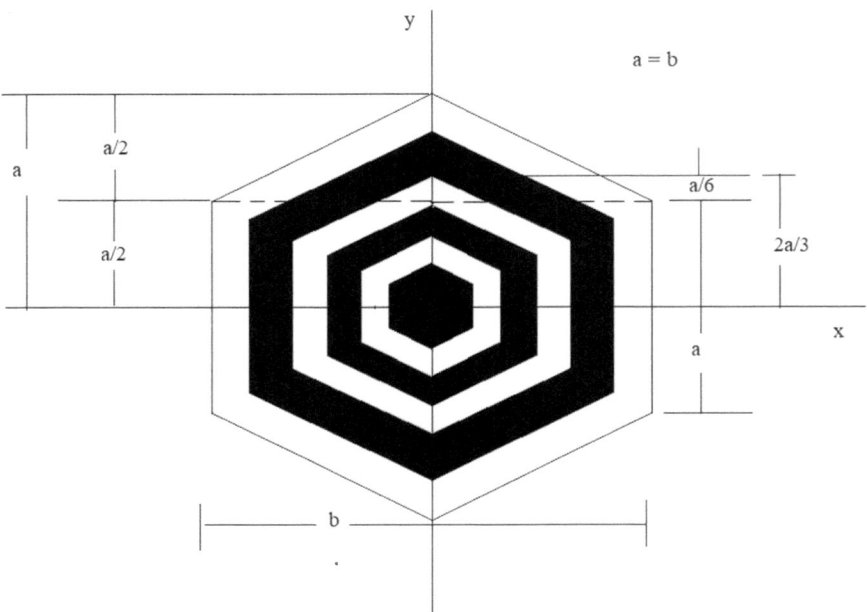

Fig. 7.1 Concentric hexagonal black and white pupil where the number of zones is $N = 6$ and the outer side length of the hexagon is a

$$P(x, y) = \sum_{i=1}^{N} \left\{ \text{rect}(\alpha_i x, \beta_i y) + \Delta\left(\alpha_i x, y - \frac{2a}{(3 \times i)}\right) + \Delta\left(x, y + \frac{2a}{(3 \times i)}\right) \right\}$$

$$- \sum_{j=2}^{N} \left\{ \text{rect}(\alpha_j x, \beta_j y) + \Delta\left(\alpha_j x, y - \frac{2a}{(3 \times j)}\right) + \Delta\left((\alpha_j x, y + \frac{2a}{(3 \times j)}\right) \right\}$$

(7.5)

where $i = 1, 3, 5, \ldots$ is odd number and $j = 2, 4, 6, \ldots$ even number.

We perform the Fourier transform on the modulated hexagonal aperture represented in Eq. (7.4), and we obtain the PSF as follows:

$$h(u, v) = \left[\frac{\sin(\pi bu)}{\pi bu}\right]\left[\frac{\sin(\pi av)}{\pi av}\right] \times \left\{1 + 2\left[\frac{\sin(\pi bu)}{\pi bu}\right]\left[\frac{\sin(\pi av)}{\pi av}\right]\cos\left[\left(\frac{4\pi a}{3\lambda f}\right)v\right]\right\}$$

$$- \left[\frac{\sin(\pi\alpha_1 bu)}{\pi\alpha_1 bu}\right]\left[\frac{\sin(\pi\beta_1 av)}{\pi\beta_1 av}\right]$$

$$\times \left\{1 + 2\left[\frac{\sin(\pi\alpha_1 bu)}{\pi\alpha_1 bu}\right]\left[\frac{\sin(\pi\beta_1 av)}{\pi\beta_1 av}\right]\cos\left[\left(\frac{2\pi}{\lambda f}\right)\alpha_1 v\right]\right\}$$

$$+ \left[\frac{\sin(\pi\alpha_2 bu)}{\pi\alpha_2 bu}\right]\left[\frac{\sin(\pi\beta_2 av)}{\pi\beta_2 av}\right]$$

$$\times \left\{1 + 2\left[\frac{\sin(\pi\alpha_2 bu)}{\pi\alpha_2 bu}\right]\left[\frac{\sin(\pi\beta_2 av)}{\pi\beta_2 av}\right]\cos\left[\left(\frac{2\pi}{\lambda f}\right)\alpha_2 v\right]\right\}$$

$$
\begin{aligned}
&- \left[\frac{\sin(\pi\alpha_3 bu)}{\pi\alpha_3 bu}\right]\left[\frac{\sin(\pi\beta_3 av)}{\pi\beta_3 av}\right] \\
&\times \left\{1 + 2\left[\frac{\sin(\pi\alpha_3 bu)}{\pi\alpha_3 bu}\right]\left[\frac{\sin(\pi\beta_3 av)}{\pi\beta_3 av}\right]\cos\left[\left(\frac{2\pi}{\lambda f}\right)\alpha_3 v\right]\right\} \\
&+ \left[\frac{\sin(\pi\alpha_4 bu)}{\pi\alpha_4 bu}\right]\left[\frac{\sin(\pi\beta_4 av)}{\pi\beta_4 av}\right] \\
&\times \left\{1 + 2\left[\frac{\sin(\pi\alpha_4 bu)}{\pi\alpha_4 bu}\right]\left[\frac{\sin(\pi\beta_4 av)}{\pi\beta_4 av}\right]\cos\left[\left(\frac{2\pi}{\lambda f}\right)\alpha_4 v\right]\right\} \\
&- \left[\frac{\sin(\pi\alpha_5 bu)}{\pi_5 bu}\right]\left[\frac{\sin(_5 av)}{\pi_5 av}\right] \\
&\times \left\{1 + 2\left[\frac{\sin(\pi\alpha_5 bu)}{\pi\alpha_5 bu}\right]\left[\frac{\sin(\pi\beta_5 av)}{\pi\beta_5 av}\right]\cos\left[\left(\frac{2\pi}{\lambda f}\right)\alpha_5 v\right]\right\}
\end{aligned}
\tag{7.6}
$$

We rewrite Eq. (7.6) in summation form as follows:

$$
\begin{aligned}
h(u, v) &= \left[\frac{\sin(\pi bu)}{\pi bu}\right]\left[\frac{\sin(\pi av)}{\pi av}\right] \times \left\{1 + 2\left[\frac{\sin(\pi bu)}{\pi bu}\right]\left[\frac{\sin(\pi av)}{\pi av}\right]\cos\left[\left(\frac{4\pi a}{3\lambda f}\right)v\right]\right\} \\
&+ \sum_{i=1}^{N-1}(-1)^i\left[\frac{\sin(\pi\alpha_i bu)}{\pi\alpha_i bu}\right]\left[\frac{\sin(\pi\beta_i av)}{\pi\beta_i av}\right] \\
&\times \left\{1 + 2\left[\frac{\sin(\pi\alpha_i bu)}{\pi\alpha_i bu}\right]\left[\frac{\sin(\pi\beta_i av)}{\pi\beta_i av}\right]\cos\left[\left(\frac{2\pi}{\lambda f}\right)\alpha_i v\right]\right\}
\end{aligned}
\tag{7.7}
$$

where $i = 1, 2, 3, ..., N - 1$ for $N = 6$.

7.2.1 Speckle Formation Using Modulated Hexagonal Apertures Combined with a Diffuser

In this section, we obtain speckle images using a concentric B/W hexagonal aperture followed by a diffuser. A hexagonal diffuser was added to obtain the results.

The laser beam is rendered parallel using spatial filtering and then incident upon the aperture as described by Eq. (7.4) followed by a randomly distributed function representing the diffuser. Hence, the complex amplitude transmitted from the cascaded aperture and the diffuser is written as follows:

$$
T(x, y) = P(x, y).d(x, y)
\tag{7.8}
$$

The diffuser is described by the following randomly distributed function:

$$
d(x, y) = \exp\{2\pi i. \operatorname{rand}(x, y)\}
\tag{7.9}
$$

The obtained speckle image is obtained in the focal plane of a converging lens in the (u, v) plane by applying the Fourier transform to Eq. (7.8) to obtain:

$$
\tilde{T}(u, v) = h(u, v) \otimes \tilde{d}(u, v)
\tag{7.10}
$$

$h(u, v)$ is analytically computed in Eq. (7.7) by operating the FT on aperture $P(x, y)$ and $\tilde{d}(u, v)$ is the FT corresponding to the diffuser function described in Eq. (7.9).

When the modulated hexagonal aperture is replaced by a uniform circular aperture, the speckle image becomes:

$$\tilde{T}(u, v) = \frac{2J_1(\alpha\rho)}{(\alpha\rho)} \otimes \tilde{d}(u, v) \tag{7.11}$$

$\alpha = \frac{2\pi r_0}{\lambda f}$, r_0 is the aperture radius and ρ is the radial coordinate in the Fourier plane (u, v).

Hence, the resulting speckle image is dependent on the aperture geometry as expected.

7.3 Results and Discussion

We fabricated a B/W hexagonal pupil where the number of zones was $N = 6$ and the corresponding line plot was $y = 256$ pixels as shown in Fig. 7.2. The aperture is resized in a 512×512 pixel matrix, and the outer side length is $a = 126$ pixels, while the internal black hexagon has a side length equal to $a/6 = 21$ pixels.

We computed the PSF by applying the fast Fourier transform (FFT) to the B/W hexagonal aperture. Hence, the B/W hexagonal apertures of the six zones and the corresponding normalized PSF are plotted in Fig. 7.3a. The total bandwidth of the central peak $= 4$ pixels. Similarly, we computed the PSF for the uniform hexagonal aperture giving the total bandwidth of the central peak $= 6$ pixels as shown in Fig. 7.3. Consequently, we showed an improved PSF in the B/W hexagonal aperture compared with that corresponding to the transparent hexagonal aperture;

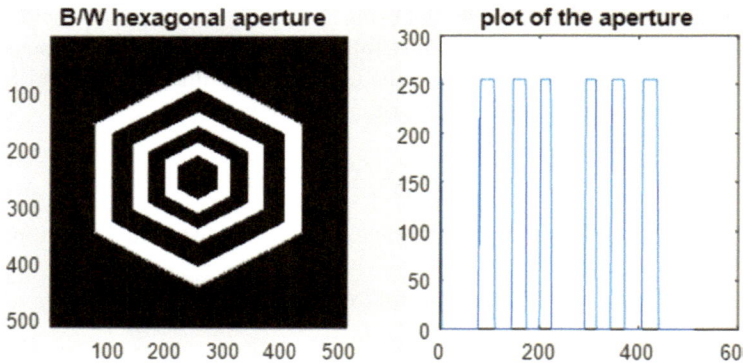

Fig. 7.2 Image is a B/W hexagonal pupil where the number of zones is $N = 6$ and the corresponding line plot is $y = 256$ pixels

hence, improved resolution is attained with the B/W hexagonal aperture. The units used in all the figures are 512×512 pixels.

We made another comparison with the uniform circular aperture of radius = 128 pixels, and we plotted the normalized PSF in Fig. 7.3c. The total bandwidth of the central peak = 6 pixels.

We observed a central peak with a small full width = 4 pixels surrounded by secondary legs with irregular triangular shapes Fig. 7.3a. Hence, we obtain improved resolution in the case of a B/W hexagonal aperture.

This irregular triangular shape is attributed to the geometry of the aperture which is composed of six triangles for a hexagonal aperture. The legs oscillate between higher and lower values with attenuation due to the B/W geometry.

We plotted the autocorrelation corresponding to the B/W hexagonal aperture compared with that corresponding to the uniform circular aperture. The autocorrelation bandwidth is two times the maximum diameter of the apertures. The lower plot shows the autocorrelation corresponding to the coronary artery image. All the plots are grouped as in Fig. 7.4.

As shown in Fig. 7.4, the autocorrelation distribution is dependent on the geometry. The B/W hexagonal aperture has fringing decay compared with the smooth decay for the uniform hexagonal pupil. The autocorrelation for the coronary arteries has a distribution specific to its geometry.

The speckle image is computed from Eq. (7.10) and plotted as in Fig. 7.5.

7.3.1 Construction of a Digital Hexagonal Diffuser and Speckle Formation

We constructed a hexagonal diffuser with dimensions of 1024×1024 pixels. There are 512 hexagonal grains where the total number of grains is 1024 and the black ratio = 0.5 as shown in Fig. 7.6a. A magnified image of the segment from the hexagonal diffuser is shown in Fig. 7.6b. It has 140 hexagonal grains. The magnified image is nearly a quarter of the image shown in Fig. 1.1a. The hexagonal diffuser is compared with the circular diffuser of the same dimensions 1024×1024 pixels, where the black ratio = 0.5 as shown in Fig. 7.6c.

A speckle image of the hexagonal diffuser with an aperture radius = 64 pixels is shown in Fig. 7.7a. The image has dimensions of 256×256 pixels.

The number of grains found in the image is 322, and the average grain size is 33.89 pixels according to the MATLAB code [1] for the image shown in Fig. 7.7b for the speckle image in Fig. 7.7a. A comparative speckle image corresponding to the circular diffuser is shown in Fig. 7.7c. When the number of grains in the image is 249, the average grain size is 34.60 pixels as shown in Fig. 7.7d.

A plot corresponding to the speckle image for the circular diffuser is shown in A while B plot for the speckle image corresponding to the hexagonal diffuser is shown in Fig. 7.8a. For the same aperture radius = 64 pixels in both plots, we found that

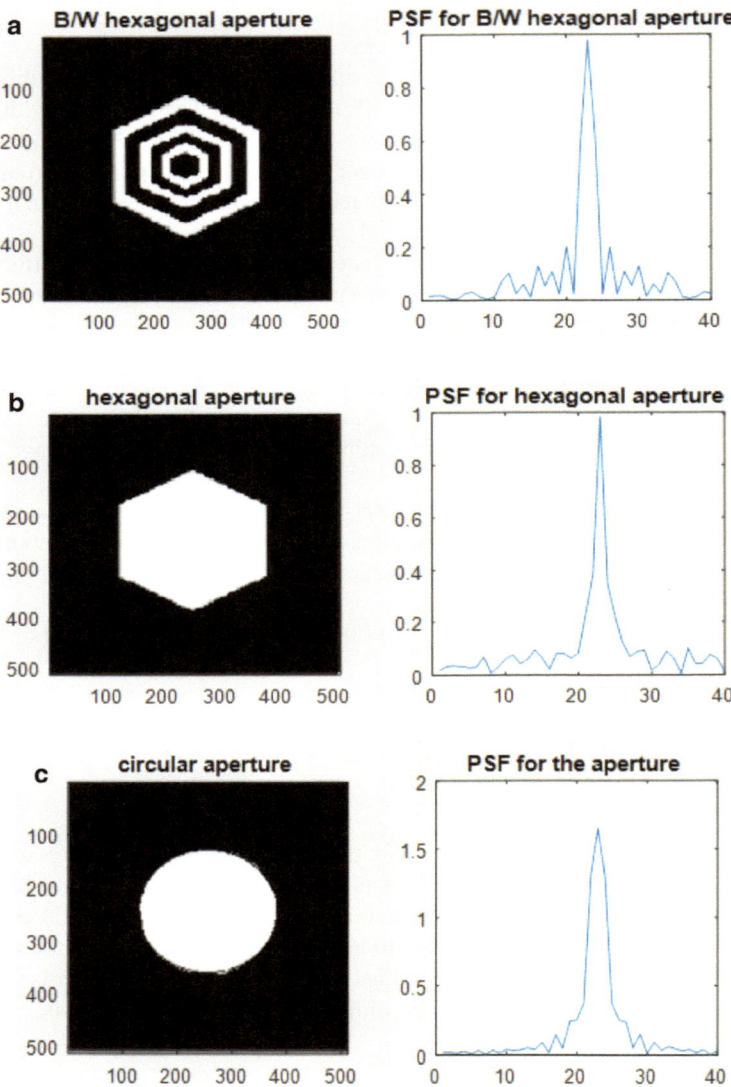

Fig. 7.3 **a** Black and white concentric hexagonal aperture of six zones and the corresponding normalized PSF. The total bandwidth of the central peak = 4 pixels. The matrix for the B/W hexagonal aperture has 512×512 pixels. **b** Uniform hexagonal aperture and the corresponding normalized PSF. The total width of the central peak = 6 pixels. **c** Uniform circular aperture with a radius = 128 pixels and the corresponding normalized PSF. The total bandwidth of the central peak is equal to 6 pixels for equal diameters as shown in Fig. 7.3b

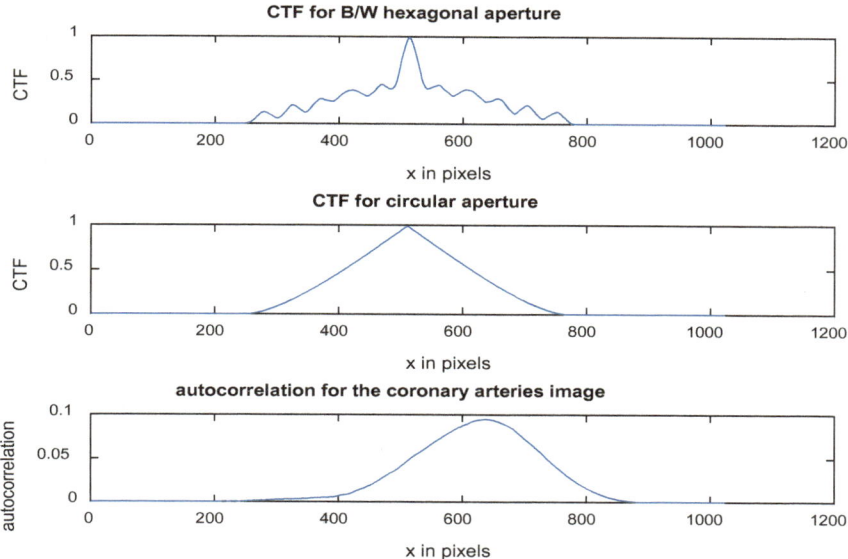

Fig. 7.4 We plot the autocorrelation corresponding to the B/W hexagonal aperture compared with that corresponding to the uniform circular aperture. The autocorrelation bandwidth is two times the maximum diameter of the apertures. The lower plot shows the autocorrelation corresponding to the coronary artery image

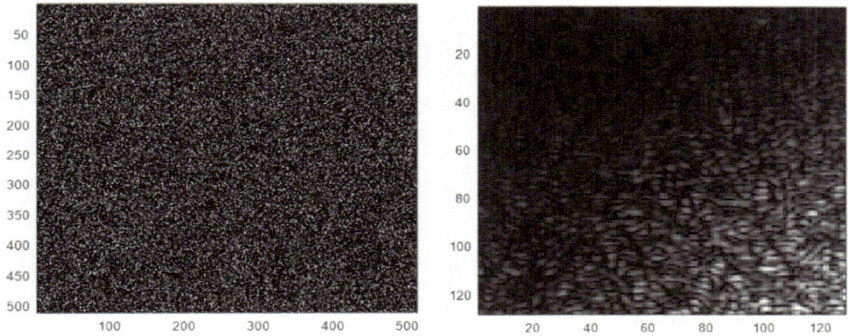

Fig. 7.5 Diffuser used in the formation of the speckle image and the corresponding speckle image using the concentric B/W hexagonal aperture shown in Fig. 7.3a and the diffuser with dimensions 512×512 pixels. The speckle image has dimensions of 512×512 pixels

the profiles are dependent on the geometry of the diffuser. We computed the speckle contrast in the case of the hexagonal diffuser, which is equal to 0.3017, and compared it with that of the circular diffuser which is 0.2588.

Fig. 7.6 **a** Hexagonal
diffuser with dimensions of
1024 × 1024 pixels, where
the black ratio = 0.5. There
are 512 hexagonal grains, for
which the total number of
grains is 1024. **b** Magnified
image of the segment from
the hexagonal diffuser is
shown in Fig. 7.6a. It has 140
hexagonal grains. **c** Circular
diffuser with dimensions of
1024 × 1024 pixels, where
the black ratio = 0.5

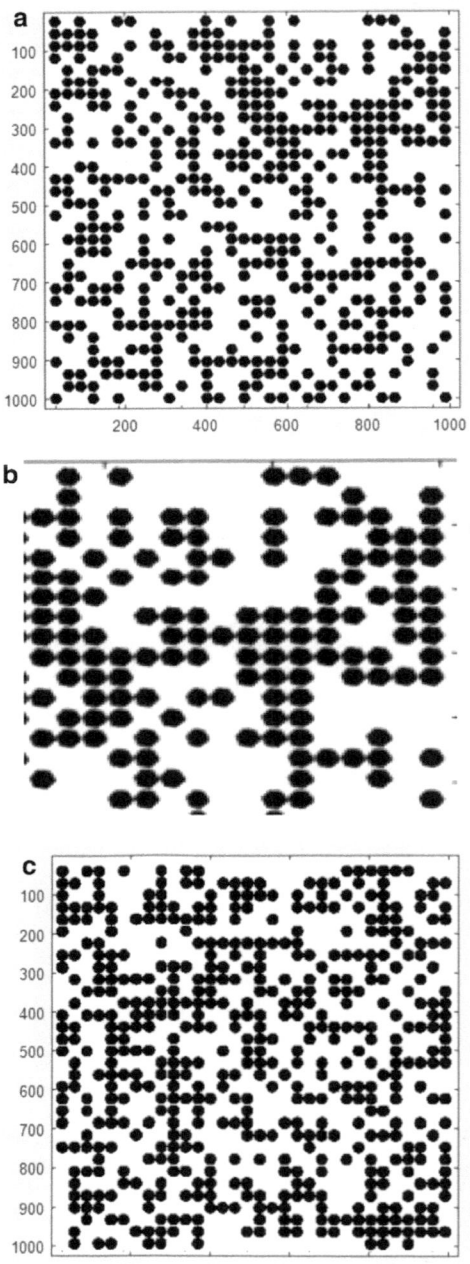

Fig. 7.7 **a** Speckle image for the hexagonal diffuser where the aperture radius = 64 pixels. The image has dimensions of 256 × 256 pixels. **b** Number of grains found in the image = 309. The average grain size is 33.89 pixels. **c** Speckle image for circular diffusers where the aperture radius = 64 pixels. The image has dimensions of 256 × 256 pixels. **d** Number of grains found in the image = 244. The average grain size is 34.60 pixels

Fig. 7.7 (continued)

We plot the speckle image produced from the magnified zone of the hexagonal diffuser in Fig. 7.8b. The speckle contrast for the magnified segment from the hexagonal diffuser is computed to be 0.2658.

7.4 Conclusions

First, we computed the PSF for concentric black and white hexagonal apertures and found that the PSF for the central lobe was lower than that for the transparent hexagonal aperture. In addition, the diffracted legs have a sharp triangular shape. We compared the PSF with that corresponding to a uniform circular aperture with curved legs depending on the aperture geometry. In addition, the legs in the diffraction pattern corresponding to the concentric B/W hexagonal aperture are strengthened compared with those corresponding to the hexagonal apertures in Fig. 7.3b and the uniform circular aperture in Fig. 7.3c.

If the central lobe starts with white in the B/W concentric hexagonal aperture the resolution decreases as expected. In contrast, the resolution is improved by obstructing the central zone as shown earlier in the famous PSF corresponding to the annular aperture than in the case of the transparent circular aperture.

The coherent transfer function (CTF) or the autocorrelation corresponding to the B/W concentric hexagonal aperture has triangular decaying fringing compared with that corresponding to the uniform hexagonal and circular apertures.

The number of grains in the speckle image of the hexagonal diffuser is greater than that corresponding to the speckle using the diffuser of circular grains. The speckle images in both cases are obtained using a circular aperture with a constant radius = 64 pixels as shown in Fig. 7.7.

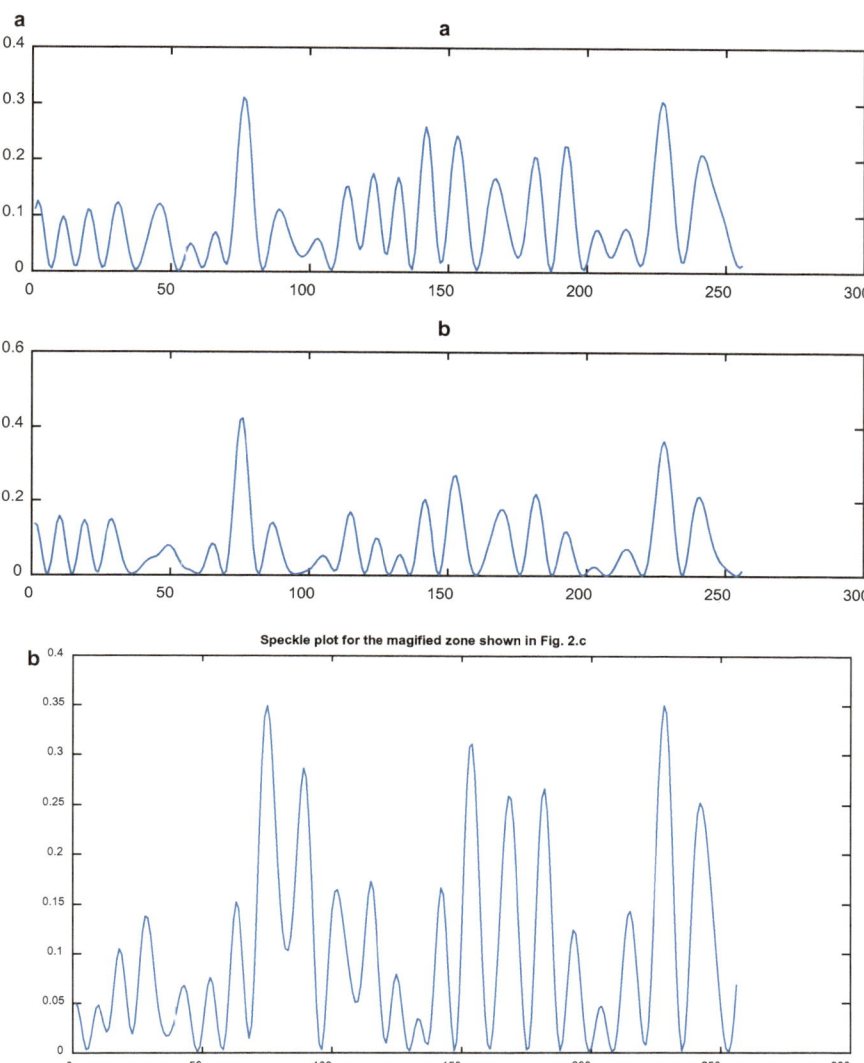

Fig. 7.8 **a** In A, we show the plot corresponding to the speckle image for the circular diffuser while in B, the plot for the speckle image corresponding to the hexagonal diffuser. **b** The plot of the speckle image corresponding to the magnified zone from the hexagonal diffuser. The speckle contrast $= 0.2658$

We showed an enhanced speckle contrast when using the hexagonal diffuser compared with the speckle contrast when using the circular grains from the diffuser.

Finally, the average grain size in the speckle image is dependent on the geometry of the diffuser. It is equal to 33.89 pixels in the case of the hexagonal diffuser, whereas it is 34.60 pixels in the case of the circular grains of the diffuser.

References

1. J.W. Goodman, in *Introduction to Fourier Optics* (Roberts and Company Publishers, Englewood, 2005)
2. J.J. Clair, A.M. Hamed, Theoretical studies on the optical coherent microscope. Optik **64**, 133–141 (1983)
3. A.M. Hamed, J.J. Clair, Image, and super resolution in optical coherent microscopes. Optik **64**, 277–284 (1983)
4. A.M. Hamed, J.J. Clair, Studies on optical properties of confocal scanning optical microscope using pupils with radially transmission distribution. Optik **65**, 209–218 (1983)
5. A.M. Hamed, Formation of speckle images formed for diffusers illuminated by modulated apertures (circular obstruction). J. Mod. Opt. **56**, 1633–1642 (2009). https://doi.org/10.1080/09500340903277792
6. A.M. Hamed, Numerical speckle images formed by diffusers using modulated conical and linear apertures. J. Mod. Opt.Mod. Opt. **56**(10), 1174–1181 (2009). https://doi.org/10.1080/09500340902985379
7. A.M. Hamed, Discrimination between speckle images using diffusers modulated by some deformed apertures: simulations. J. Opt. Eng. **50**(1), 018202 (2011). https://doi.org/10.1117/1.3530085
8. A.M. Hamed, Study of graded index and truncated apertures using speckle images. Precis. Instrument. Mech. PIM **3**, 144–152 (2014)
9. A.M. Hamed, Improvement of point spread function (PSF) using linear-quadratic aperture. Optik **131**, 838–849 (2017). https://doi.org/10.1016/2016.11.201
10. R. HeintZman, T. Huser, Super resolution structured illumination microscopy. Chem. Rev. **117**(23), 13890–13908 (2017) https://doi.org/10.1021/acs.chemrev.7b00218
11. S. Krishnendu, J. Joby, An overview of structured illumination microscopy: recent advances and perspectives. J. Opt. **23** (2021) https://doi.org/10.1088/2040-8986/ac3675
12. Q. Jian, H. Xu, M. Ang, et al., Stochastic optical reconstruction microscopy. Curr. Protocol Cytom. **81** (2017). https://doi.org/10.1002/cpcy.23
13. C.J.R. Sheppard, The development of microscopy for super-resolution. Appl. Sci. **11**, 8981 (2021). https://doi.org/10.3390/11198981
14. A.M. Hamed, Application of a hexagonal aperture on the confocal scanning laser microscope. Opt. Quant. Electron.Electron. **55**, 749 (2023). https://doi.org/10.1007/s11082-023-04920-8
15. S. Itoh, T. Matsuo et al., Point spread function of hexagonally segmented telescopes by new symmetrical formulation. MNRAS **483**, 119–131 (2019)
16. A.M. Mohamed, S.H. Mohsen, et al.: A hexagonal aperture to reduce spherical aberration. Neuro Quantol. **20**(6), 2599 (2022)
17. E. Sabatke, J. Burge et al., Analytic diffraction analysis of a 32-m telescope with hexagonal segments for high-contrast imaging. Appl. Opt. **44**(8), 1360–1365 (2005)
18. A. Deng, Y. Zheng et al., Improved spatial resolution using focal modulation microscopy with a Tai Chi aperture. Opt. Express **29**(12), 18263–18276 (2021). https://doi.org/10.1364/OE.426600
19. N. Chen, C.H. Wong, C.J.R. Sheppard, Focal modulation microscopy. Opt. Express **16**(23), 18764–18769 (2008)

20. W. Gong, K. Si, N. Chen, C.J.R. Sheppard, Improved spatial resolution in fluorescence focal modulation microscopy. Opt. Lett. **34**(22), 3508–3510 (2009). https://doi.org/10.1364/OL.34. 003508
21. C. Wu, Y. Zheng et al., Improvements with divided cosine-shaped apertures in confocal microscopy. Opt. Commun.Commun. **442**, 71–76 (2019)
22. K. Si, W. Gong, N. Chen, C.J.R. Sheppard, Enhanced background rejection in thick tissue using focal modulation microscopy with quadrant apertures. Opt. Commun. **284**(5), 1475–1480 (2011)

Chapter 8
Investigation of Irregular Apertures and Applications in Speckle Imaging

We investigated an irregular aperture, and we calculated the impulse response corresponding to this aperture. In addition, we investigated the impulse response in the case of rotated irregular apertures.

Second, we differentiate speckle images using different orientations of the irregular aperture. Then, we reconstructed the irregular aperture at different orientations from the speckle pattern.

8.1 Introduction

There are many references for computing the point spread function (PSF) or the impulse response corresponding to regular circular and modulated apertures [1–8]. Recently, an algorithm was presented in [9] for measuring the PSF from aperture images of arbitrary shapes. Fraunhofer diffraction of irregular apertures by the Heisenberg uncertainty Monte Carlo (HUMC) model was reported in [10]. They showed that the diffracted intensity distributions of simple apertures obtained by the HUMC model are in good agreement with the results calculated from analytical expressions. Others presented a method for numerically calculating the point spread function of an irregular aperture objective lens [11].

In this chapter, we computed the impulse response using irregular apertures. In addition, we present an application for speckle imaging of irregular apertures. We reconstructed the aperture affected by the diffuser as a noise. Finally, we discuss the results and present the conclusion.

© The Author(s), under exclusive license to Springer Nature Switzerland AG 2024   91
A. Hamed, *Speckle Imaging Using Aperture Modulation*,
SpringerBriefs in Applied Sciences and Technology,
https://doi.org/10.1007/978-3-031-58300-1_8

8.2 Method

The confocal scanning laser microscope used is described in Fig. 8.1. The coherent laser beam is spatially filtered and rendered parallel. The beam is incident upon confocal microscope objectives obstructed by irregular apertures. The image of the detection plane (x_2, y_2) is the result of the convolution product of the resulting impulse response and the object placed in the plane (x, y). The convolution is realized from the mechanical scanning of the object and the electronic scanning in the detection plane. The resulting impulse response is the product of the impulse response corresponding to each objective of the confocal microscope [12–15].

We fabricated an aperture with an irregular shape and matrix dimensions of 1024 × 1024 pixels. Then, we computed the Fourier spectrum image corresponding to the irregular aperture. We computed the coherent transfer function (CTF) considering two symmetric objectives provided with irregular apertures using the CSLM. In addition, we computed the CTF in the case of one unrotated aperture, while the other was rotated by angles of 30, 60, 90, …, 180°.

We apply the irregular aperture in the formation of a speckle pattern resulting from the Fourier spectrum of the diffuser multiplied by the Fourier spectrum of the aperture. The resulting Fourier spectrum or speckle pattern is formed in the focal plane of a converging lens. The irregular aperture is reconstructed by operating the FFT on the speckle pattern.

8.2.1 Effect of Irregular Aperture Rotation on the Formed Speckle Pattern

We assumed that the collimated laser beam incident on the diffuser was obstructed by an irregular aperture placed in the plane (x, y). The transmitted complex amplitude is represented as follows:

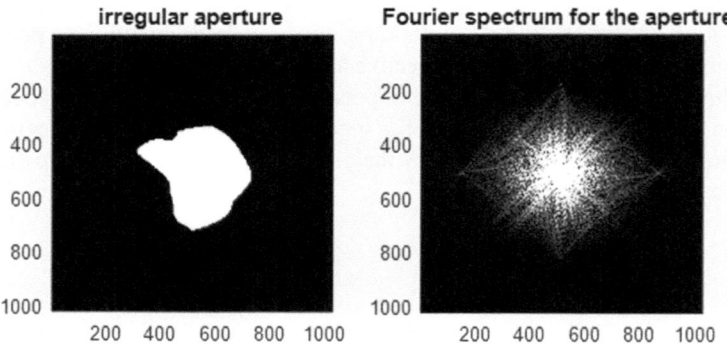

Fig. 8.1 On the left-hand side, an aperture of irregular shape is shown. The Fourier spectrum is shown on the right-hand side. The matrix dimensions are 1024 × 1024 pixels

$$A(x, y) = P(x, y).d(x, y) \tag{8.1}$$

where $d(x, y)$ is the complex amplitude of the diffuser represented by a randomly distributed function and $P(x, y)$ represents the irregular aperture transmittance.

Then, using a converging lens of focal lens f, we apply the Fourier transform to Eq. (8.1) to obtain the Fraunhofer diffraction in the (u, v) plane as follows:

$$B(u, v) = \iint\limits_{-\infty}^{\infty} A(x, y) \exp\left\{-\frac{j2\pi}{\lambda f}(ux + vy)\right\} dxdy \tag{8.2}$$

We rewrite Eq. (8.2) as follows:

$$B(u, v) = F.T.\{A(x, y)\} = F.T.\{P(x, y).d(x, y)\} \tag{8.3}$$

We perform the transformation and make use of convolution and Fourier transform properties and we obtain the following:

$$B(u, v) = h(u, v) \otimes s(u, v) \tag{8.4}$$

$h(u, v)$ is the F.T. corresponding to the irregular aperture, while $s(u, v)$ is the complex amplitude of the obtained speckle pattern for $h(u, v) = \delta(u, v)$. In reality, the speckle pattern is affected by the F.T. corresponding to the aperture named the point spread function (PSF) [7]. Hence, the speckle intensity formed in the (u, v) plane is the modulus square of $B(u, v)$. Hence, we write the speckle image corresponding to the unrotated irregular aperture as follows:

$$I(u, v) = h(u, v) \otimes s(u, v)^2 \tag{8.5}$$

If we incline the irregular aperture by a certain angle α, we represent it as follows:

$$P_{\text{rot.}}(x', y') = P(x, y).\exp\left\{\frac{j2\pi}{\lambda} x \cos(\alpha)\right\}$$

We computed the PSF corresponding to the rotated aperture $h'(u, v)$ as follows:

$$h'^{(u,v)} = F.T.\left\{P(x, y).\exp\left[\frac{j2\pi}{\lambda} x \cos(\alpha)\right]\right\} \tag{8.6}$$

As before, we make use of the properties of the Fourier transform and convolution operation, and we easily obtain:

$$h'(u, v) = h(u, v) \otimes \delta[u - f \cos(\alpha)] = h(u - f \cos(\alpha), v) \tag{8.7}$$

Hence, following the same analysis, we get the speckle image shifted by an amount of $f \cos(\alpha)$ along the direction, where u corresponds to the inclination angle assumed in the x direction. Hence Eq. (8.5) for the inclined aperture becomes:

$$I(u, v) = |h(u - f \cos(\alpha), v) \otimes s(u, v)|^2 \tag{8.8}$$

Consequently, we obtain a shifted speckle pattern along the u direction, while in general, the shift is dependent on the direction of inclination.

In the other case where the irregular aperture rotates in its plane with an angle of α, the speckle image is affected by the rotation of the irregular aperture. The pupil $P_{\text{rot.}}(x', y')$ has rotated axes (x', y'), where (x', y') are given in terms of the original coordinates (x, y) as follows:

$x' = x \cos \alpha + y \sin \alpha$ and $y' = y \cos \alpha - x \sin \alpha$. Hence, the speckle image is dependent on the rotated irregular aperture.

8.3 Results

On the left-hand side, an irregular shape is shown. The Fourier spectrum is shown on the right-hand side. The matrix dimensions corresponding to both images are 1024 × 1024 pixels.

We plot the Fourier spectrum corresponding to the irregular aperture in Fig. 8.1. The plots are at $y = 512$ and $x = 512$ pixels as shown in Fig. 8.2.

The effect of rotation of the irregular apertures is summarized in six images as shown in Fig. 8.3. We showed five rotated irregular apertures at angles of 30, 60, 90, 120, and 150° and compared them with the unrotated aperture. All the images have dimensions of 1024 × 1024 pixels.

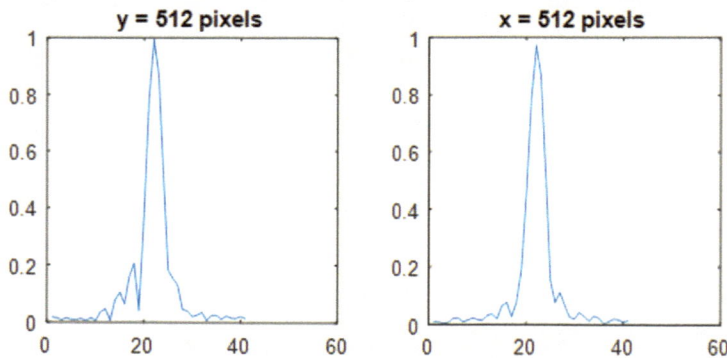

Fig. 8.2 Plot of the Fourier spectrum corresponding to the irregular aperture in Fig. 8.1 is at $y = x = 512$ pixels. We considered the central region in the range from 492 to 532 pixels

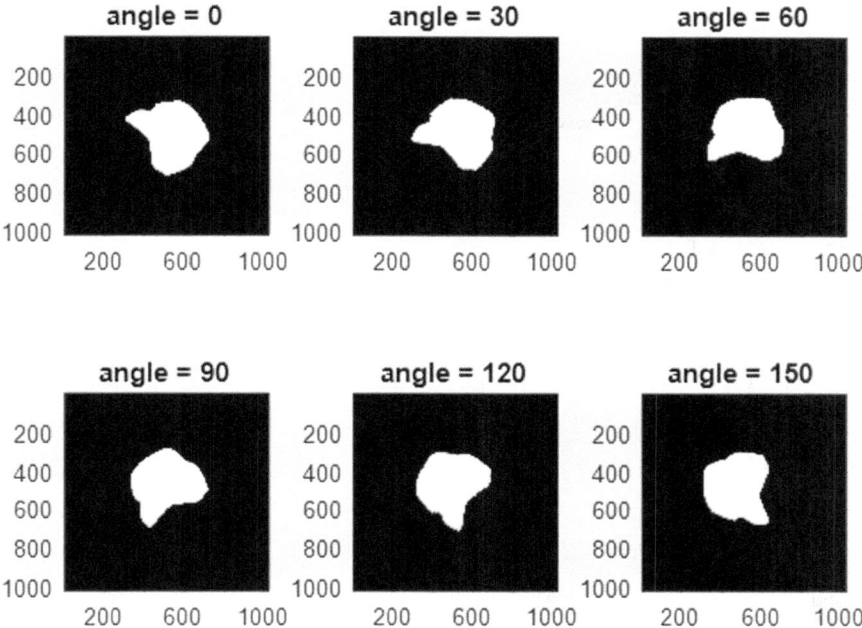

Fig. 8.3 Six images of rotated irregular apertures at angles of 30, 60, 90, 120, and 150° are shown and compared with the unrotated aperture

We calculated the Fourier spectrum image using the fast Fourier transform (FFT) and we computed the corresponding plot at $y = 512$ pixels considering the unrotated aperture as shown in Fig. 8.4a. We computed the Fourier spectrum image and the corresponding plots at $y = 512$ pixels for the rotated apertures at angles of 30 and 60° as shown in the Fig. 8.4b, c. We considered the central region in the range from 482 to 542 pixels for all plots in Fig. 8.4a–c.

We computed the coherent transfer function (CTF) in the case of two symmetric irregular apertures for both microscope objectives and its plot at $y = 512$ pixels considering the CSLM. In this case, the CTF is the autocorrelation corresponding to the two symmetric irregular apertures as shown in Fig. 8.5a.

We calculated the cross-correlation or the coherent transfer function (CTF) in the case of one unrotated aperture and the other rotated with an angle of 30° as shown in Fig. 8.5b. The corresponding plot at $y = 512$ pixels is shown in the same figure. We obtained other CTF image and corresponding plots as shown in Fig. 8.5c.

We investigated the speckle images produced by the diffuser obstructed by an irregular aperture formed in the focal plane of the converging lens. We obtained three different speckle images corresponding to an unrotated irregular aperture as shown in Fig. 8.6a and other speckle images for the rotated aperture at a high angle of $\alpha = 30°$, and very small angle of $\alpha = 1°$ as shown in Fig. 8.6b, c.

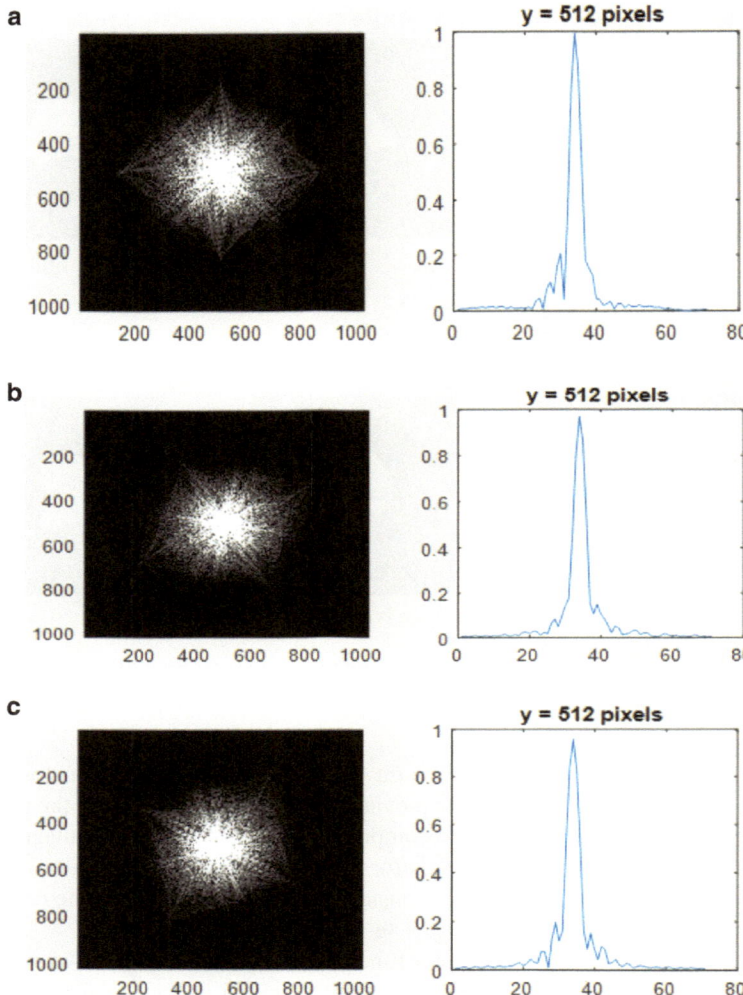

Fig. 8.4 **a** Image of the Fourier spectrum and its normalized plot for the unrotated aperture is shown. We considered the central region in the range from 482 to 542 pixels. **b** Image of the Fourier spectrum and its normalized plot corresponding to the rotated aperture at an angle $= 30°$ is shown. The range is from 482 to 542 pixels around the center. **c** Image of the Fourier spectrum and its normalized plot corresponding to the rotated aperture at an angle $= 60°$ is shown. The range is from 482 to 542 pixels around the center

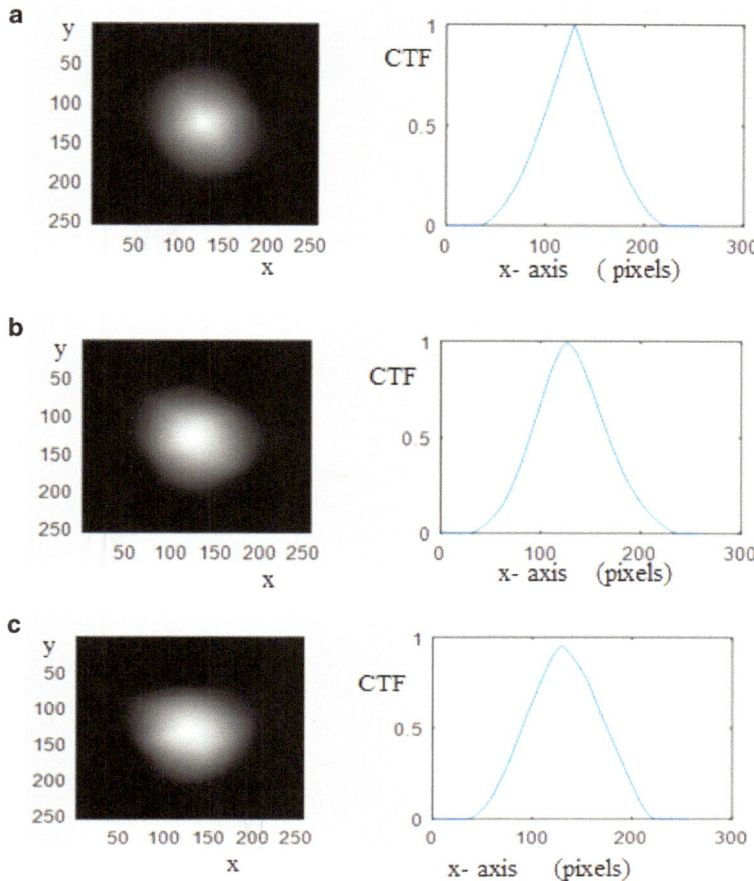

Fig. 8.5 a Autocorrelation image or coherent transfer function (CTF) for both microscope objectives and plot of two symmetric irregular apertures. We considered the CSLM. **b** Cross-correlation image or coherent transfer function (CTF) for one unrotated aperture and the other rotated at an angle of 30° and its plot. **c** Cross-correlation image or coherent transfer function (CTF) obtained using CSLM in the case of one unrotated aperture and the other rotated by an angle 60° and its plot

We show speckle image plots corresponding to the unrotated and rotated irregular aperture in Fig. 8.7a where a small angle of rotation is assumed to be $\alpha = 1°$. In Fig. 8.7b, we show speckle image plots corresponding to the unrotated and rotated irregular aperture where the angle of rotation is assumed to be $\alpha = 30°$. In both plots shown in Fig. 8.7a, b, we considered the plotted zone in the range from 44 to 84 pixels where the total range is 256 pixels as shown in Fig. 8.6.

We obtained irregular apertures that were either rotated or not rotated by operating the FFT upon the complex amplitude corresponding to the speckle pattern and we represented them as shown in Fig. 8.7a–c.

Fig. 8.6 **a** Speckle image using a diffuser obstructed with an irregular aperture where the angle of rotation α = 0. The diffuser has dimensions of 1024×1024 pixels. **b** Speckle image using a diffuser obstructed with the rotated irregular aperture where the angle of rotation $\alpha = 30°$. The diffuser has dimensions of 1024×1024 pixels. **c** Speckle image using a diffuser obstructed with the rotated irregular aperture where the angle of rotation is very small $\alpha = 1°$. The diffuser has dimensions of 1024×1024 pixels

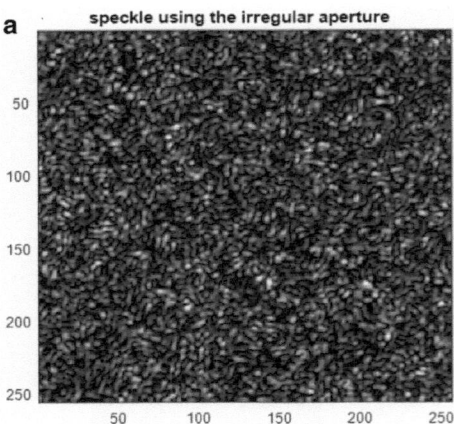

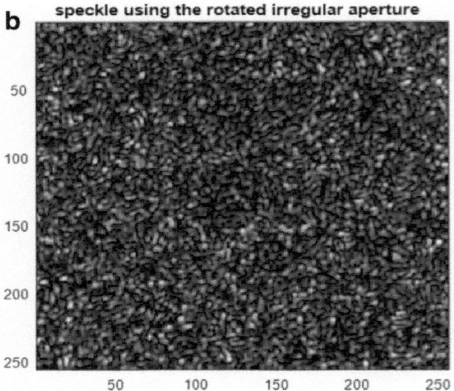

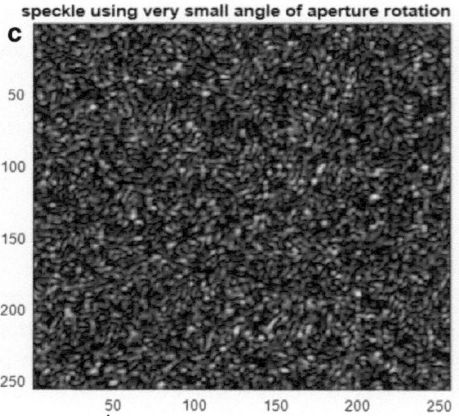

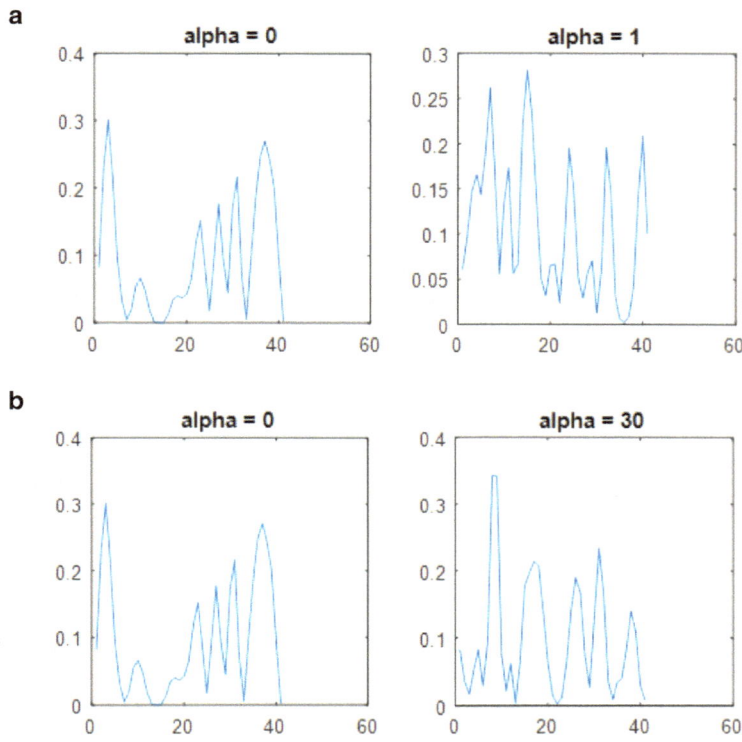

Fig. 8.7 **a** Speckle image plots corresponding to the unrotated and rotated irregular aperture. A small angle of rotation is assumed to be $\alpha = 1°$. We considered the plotted zone in the range from 44 to 84 pixels where the total range was 256 pixels as shown in Fig. 8.6. **b** Speckle image plots corresponding to the unrotated and rotated irregular aperture. The angle of rotation is assumed to be $\alpha = 30°$. We considered the plotted zone in the range from 44 to 84 pixels where the total range was 256 pixels as shown in Fig. 8.6

8.4 Discussion

Figure 8.1 shows that the diffraction pattern or the Fourier spectrum corresponding to the irregular aperture has many diffracted legs. The difference between these legs was observed by the naked eye and we confirmed by plotting the difference in the orthogonal coordinates as shown in Fig. 8.2 or by rotating the irregular aperture and computing the corresponding Fourier spectrum in a certain direction as shown in Fig. 8.4a–e.

As shown in Fig. 8.5a, we obtained a sharp autocorrelation image for two symmetric irregular apertures for both microscope objectives. This represents the coherent transfer function (CTF) in the CSLM. The combination of unrotated and rotated irregular apertures in Fig. 8.5b–e results in a different peak curved around the center of the CTF. In addition, compared with those of the autocorrelation curve, all cross-correlation curves had different widths. Hence, the CTF is dependent on the orientation of the irregular apertures in the planes.

We discriminated between irregular apertures at different orientations as shown in Fig. 8.6a–c. Referring to the plots shown in Fig. 8.7a, b, we obtained different shapes for the same diffuser since the rotation of the irregular aperture affects the shape of the plot as expected. The speckle pattern which has random constructive and destructive interference is sensitive to small angles of aperture rotation at $\alpha = 1°$. Hence, we deduced that the speckle imaging produced by a diffuser obstructed by an irregular aperture is sensitive to rotation. We can use this technique for aperture alignment in optical systems for CLSM.

8.5   Conclusion

The Fourier spectrum corresponding to the irregular apertures is affected by the angle of aperture rotation as shown in the plots in Fig. 8.4. In addition, the autocorrelation or the CTF corresponding to the irregular aperture is different from the cross-correlation or the CTF using one unrotated aperture and the other rotated at a certain angle as shown in the plots in Fig. 8.5.

We showed that the aperture rotation affected the speckle pattern formed using a certain diffuser. In addition, the speckle pattern is sensitive to the irregular aperture rotation for a small angle of rotation $\alpha = 1°$. Hence, we can identify the aperture rotation from the speckle image as shown in Figs. 8.6 and 8.7. Finally, we reconstructed the irregular aperture from the speckle images.

References

1. A.M. Hamed, Formation of speckle images formed for diffusers illuminated by modulated apertures (circular obstruction). J. Mod. Opt. **56**, 1633 (2009). https://doi.org/10.1080/095003 40903277792
2. A.M. Hamed, Numerical speckle images formed by diffusers using modulated conical and linear apertures. J. Mod. Opt.Mod. Opt. **56**, 1174 (2009). https://doi.org/10.1080/095003409 02985379
3. A.M. Hamed, Improvement of point spread function (PSF) using linear-quadratic aperture. Optik **131**, 838 (2017). https://doi.org/10.1016/j.ijleo.2016.11.201
4. A.M. Hamed, Discrimination between speckle images using diffusers modulated by some deformed apertures. Opt. Eng. **50**, 1 (2011). https://doi.org/10.1117/1.3530085
5. A.M. Hamed, T.A. Al-Saeed, J. Mod. Opt. **62**, 801 (2015)
6. A.M. Hamed, Image processing of Ramses II statue using speckle photography modulated by a new Hamming- Linear aperture. Pram. J. Phys. **94**, 126 (2020)
7. A.M. Hamed, Contrast of laser speckle images using some modulated apertures. Pram J. Phys. **95**, 122 (2021)
8. A.M. Hamed, Speckle imaging of annular Hermite Gaussian laser beam. Pram. J. Phys. **95**, 202 (2021)

9. P.N.H. Nakashima, A.W.S. Johnson, Measuring the PSF from aperture images of arbitrary shape—an algorithm. Ultramicroscopy **94**, 135–148 (2003)
10. Y. Yuan, R. Kuan, et al., Fraunhofer diffraction of irregular apertures by Heisenberg uncertainty Monte Carlo model. Particuology **24**, 151–158 (2016)
11. X. Wu, N.D. Lai, Q. Li, A method for numerical calculation of point spread function of an irregular aperture objective lens. Phys. Scripta **93**, 8 (2018)
12. C.J.R. Sheppard, A. Choudhury, Image formation in the scanning microscope. Opt. ActaActa **24**, 1051–1073 (1977)
13. I.J. Cox, C.J.R. Sheppard, T. Wilson, Supper resolution by confocal fluorescence microscopy. Optik **60**, 391–396 (1982)
14. T. Wilson, and C.J.R. Sheppard, in *Theory and Practice of Scanning Optical Microscopy* (Academic Press, London, 1984)
15. J.B. Pawley (ed.), in *Handbook of Biological Confocal Microscopy*, 2nd edn. (Plenum Press, New York, 1995)

Chapter 9
Speckle Imaging of Annular Hermite Gaussian Laser Beam

The Fourier transform (FT) corresponding to the diffuser has a comb function realized from uniform illumination; hence the speckle image is nearly equal to the comb point spread function (PSF). Consequently, the resulting distribution permits computation of the speckle size from the FWHM corresponding to the PSF of the HG_{mn} annular aperture. In the second method, the numerical autocorrelation of the speckle images using new apertures is computed and the speckle sizes are obtained. The results of both methods are compared with the speckle images obtained using a circular aperture. The maximum number of spots produced in the HG_{mn} modes given by combinations $m, n \in [0, 5]$ is $(m + 1) \times (n + 1) = 36$ and all images and plots are obtained using MATLAB codes.

9.1 Introduction

When a laser beam is incident on a diffuser considered to be a rough surface, the light reflected and scattered from the surface has a granular structure called a speckle. In this way, the intensity of the scattered field will be due to the locus of bright spots, constructive interference, interlacement with dark spots, and destructive interference. Numerous studies on speckle formation using uniform illumination and Gaussian illumination of a single mode TEM0,0 beam have been presented [1–5]. Surface roughness measurements based on the intensity correlation function of scattered light were investigated in [4, 5]. Although the speckle was originally regarded as an undesired noise that should be removed for digital holography [2], it has become a powerful optical tool that has found applications in fields such as optical metrology, imaging, and medicine [6–13]. Speckle formation by higher-order Laguerre Gaussian modes is characterized by a nonhomogeneous intensity distribution, and the mean speckle size is independent of the intricate structure of the modes [14–16]. A Hermite Gaussian beam is considered in the formation of speckle images, and the

© The Author(s), under exclusive license to Springer Nature Switzerland AG 2024
A. Hamed, *Speckle Imaging Using Aperture Modulation*,
SpringerBriefs in Applied Sciences and Technology,
https://doi.org/10.1007/978-3-031-58300-1_9

speckle size is computed from the autocorrelation of speckle images [16]. In [17–19], linear Hamming apertures and other modulated apertures were considered in the formation of speckle images, additionally, the computation of the autocorrelation of speckle images was outlined in [17]. In this chapter, an annular Hermite Gaussian aperture of different transverse laser modes is suggested. The speckle formation using this new aperture is investigated using two different methods allowing speckle size computation. Finally, the results are discussed followed by the conclusion.

9.2 Theoretical Analysis

A transverse Hermite Gaussian laser beam is incident on a diffuser limited by an annular aperture, and the reflected light from the diffuser is gathered by a converging lens L in the focal plane. The diffracted intensity has a granular speckle pattern affected by the nonuniform beam. This process is mathematically described in the following analysis.

Starting from beam propagation, the complex amplitude of the laser beam or the electric field is described as follows:

$$A(r, z) = \frac{w_0}{w(z)} H_m\left[\frac{x\sqrt{2}}{w(z)}\right] H_n\left[\frac{y\sqrt{2}}{w(z)}\right] \exp\left(-\frac{r^2}{w^2(z)}\right) \exp\left[-jk\frac{r^2}{2R(z)}\right] \exp[-j\phi]$$

(9.1)

for lower-order modes, $m = 0, 1, 2$ and $n = 0, 1, 2$.

The intensity distribution of the beams at different orders is computed and represented as follows:

$$I_{0,0}(r, z) = \frac{w_0^2}{w^2(z)} \exp\left(-\frac{2r^2}{w^2(z)}\right)$$

(9.2)

$$I_{1,0}(r, z) = 8x^2 \frac{w_0^2}{w^4(z)} \exp\left(-\frac{2r^2}{w^2(z)}\right)$$

(9.3)

$$I_{2,0}(r, z) = \left[\frac{4w_0^2}{w^2(z)}\right]\left[\frac{4x^2}{w^2(z)} - 1\right]^2 \exp\left(-\frac{2r^2}{w^2(z)}\right)$$

(9.4)

$$I_{3,0}(r, z) = \left[\frac{8w_0^2 x^2}{w^4(z)}\right]\left[\frac{4x^2}{w^2(z)} - 3\right]^2 \exp\left(-\frac{2r^2}{w^2(z)}\right)$$

(9.5)

$$I_{4,0}(r, z) = \left(\frac{4w_0^2}{w^2(z)}\right)\left[16\frac{x^4}{w^4(z)} - 24\frac{x^2}{w^2(z)} + 3\right]^2 \exp\left(-\frac{2r^2}{w^2(z)}\right)$$

(9.6)

 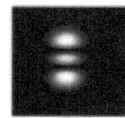 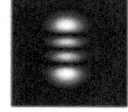 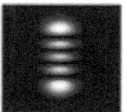 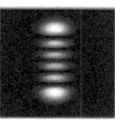

Fig. 9.1 Intensity distribution corresponding to the Hermite Gaussian laser beam for different transverse modes $HG_{n,0}$, where $n \in [1, 5]$

$$I_{5,0}(r, z) = \left(\frac{16w_0^2 x^2}{w^4(z)}\right)\left[16\frac{x^4}{w^4(z)} - 40\frac{x^2}{w^2(z)} + 15\right]^2 \exp\left(-\frac{2r^2}{w^2(z)}\right). \quad (9.7)$$

The intensity for the other modes $I_{0,1}, I_{0,2}, \ldots, I_{0,5}$ is obtained by replacing x with y in Eqs. (9.3–9.7). Similarly, $I_{1,2}, I_{2,1}, I_{1,1}, I_{2,2}$, etc. are obtained from the combinations of Eqs. (9.3) and (9.4).

The intensity distributions corresponding to the different modes are plotted in Fig. 9.1. If the Hermite Gaussian laser beam of amplitude $I_{m,n}$ is incident on the diffuser obstructed by an annular aperture, then the reflected beam is represented as follows:

$$B(x, y) = A(x, y).d(x, y). \text{annul}(\delta r) \quad (9.8)$$

where $d(x, y)$ is the complex amplitude of the diffuser and is represented as follows:

$$d(x, y) = \exp\{2\pi i \text{ rand } (x, y)\}$$

where $\text{rand}(x, y)$ represents a randomly distributed function in the diffuser plane and $\text{annul}(\delta r)$ represents the annular aperture of width $\delta r = r_{\text{ext.}} - r_{\text{int.}}$ and the annulus radius $= 0.1 \times$ total radius, e.g. $\delta r = 10$ pixels for $r_{\text{ext.}}=100$ pixels. The radial amplitude $A(r)$ is obtained by taking the squared root of Eqs. (9.2–9.7), where $r = (x, y)$ is the radial coordinate in the aperture plane.

A converging lens L of focal length f is used to gather the scattered and reflected light from the diffuser limited by the HG_{mn} annular aperture using Eq. (9.8). Hence, in the focal plane of the lens L, the speckle pattern formed in the case of the nonuniform laser beam is computed by performing the Fourier transform (FT) on Eq. (9.8) follows:

$$\tilde{B}(\rho) = FT\{A(r).d(r).\text{annul}(\delta r)\} \quad (9.9)$$

From the properties of the FT and convolution operations, the FT of the multiplication is transformed into the convolution of the FT corresponding to each function. Hence, we write Eq. (9.9) as follows:

$$\tilde{B}(\rho) = FT\{A(r)\} \otimes FT\{d(r)\} \otimes FT\{\text{annul}(\delta r)\} \quad (9.10)$$

where the symbol $\otimes$ represents the convolution operation. Solving the FT for each term gives [20]

$$FT\{\text{annul}(\delta r)\} = FT\left\{\text{cir}\left(\frac{r}{r_1}\right) - \text{cir}\left(\frac{r}{r_2}\right)\right\} = 2\left\{\frac{J_1(W)}{W} - \epsilon^2 \frac{J_1(W_1)}{W_1}\right\} \tag{9.11}$$

where $\epsilon = 0.9 = \frac{W_1}{W}$.

r_1 is the external radius of the circle, and the reduced coordinate $W = \frac{2\pi r_1 \rho}{\lambda f}$.

$\rho = (u, v)$ is the radial coordinate in the focal plane.

$$FT\{d(r)\} = \tilde{d}(\alpha\rho) \tag{9.12}$$

where $\alpha = \frac{2\pi r_1}{\lambda f}$ and r_1 is the external radius of the annulus limiting the diffuser. The FT of the Hermite Gaussian amplitude is dependent upon the m and n transverse modes.

Substituting Eqs. (9.11) and (9.12) into Eq. (9.10), we obtain:

$$\tilde{B}(\rho) = 2\left\{\tilde{A}(\rho) \otimes \tilde{d}(\alpha\rho) \otimes \left\{\frac{J_1(W)}{W} - \epsilon^2 \frac{J_1(W_1)}{W_1}\right\}\right. \tag{9.13}$$

In the case of uniform illumination, the Hermite Gaussian nonuniform illumination is replaced by a constant amplitude and hence $FT\{A(r)\} = FT\{a\} = a\delta(\rho)$.

Equation (9.13) becomes:

$$\tilde{B}(\rho) = \tilde{d}(\alpha\rho) \otimes 2\left\{\frac{J_1(W)}{W} - \epsilon^2 \frac{J_1(W_1)}{W_1}\right\} \tag{9.14}$$

In the case of an open circular aperture, the speckle pattern is represented as the convolution of the static speckle, and the PSF corresponds to the circular aperture where the central width of the diffraction pattern nearly represents the speckle size. Equation (9.14) becomes:

$$\tilde{B}(\rho) = \tilde{d}(\alpha\rho) \otimes 2\left\{\frac{J_1(W)}{W}\right\} \tag{9.15}$$

The intensity distribution of speckle images using either uniform illumination or nonuniform Hermite Gaussian illumination is the modulus square of $\tilde{B}(\rho)$. For uniform illumination a circular aperture

$$I(\rho) = \left|\tilde{d}(\alpha\rho) \otimes 2\left\{\frac{J_1(W)}{W}\right\}\right|^2 \tag{9.16}$$

For nonuniform illumination, in the case of a Hermite Gaussian laser beam, the intensity for uniform illumination without apertures is:

$$I(\rho) = |\tilde{d}(\alpha\rho) \otimes \tilde{A}(\rho)|^2 \tag{9.17}$$

The theoretical formula relating the spot size for the HG laser modes and the speckle size is [16]:

$$S = \frac{\lambda}{\mathrm{NA}}, \text{ where } \mathrm{NA} = \left(\frac{2w_0}{f}\right)[(2n+1)(2m+1)]^{1/4} \tag{9.18}$$

NA is the numerical aperture in the case of HG illumination, and f is the focal plane of the converging lens gathering the reflected and scattered light from the diffuser.

9.3 Computing the Speckle Size from the FWHM of the PSF

The complex amplitude corresponding to the speckle image obtained in Eq. (9.13), in the case of HG_{mn} annular aperture is rewritten as follows:

$$\tilde{B}(\rho) = 2\{\tilde{H}(\rho) \otimes \tilde{d}(\alpha\rho) \otimes \left\{\frac{J_1(W)}{W} - \epsilon^2 \frac{J_1(W_1)}{W_1}\right\} \tag{9.19}$$

where $\tilde{A}(\rho) = \tilde{H}_{rm}(u, v)$ and $\tilde{d}(\alpha\rho) = \tilde{d}(u, v)$.

For a parallel beam of uniform illumination in the absence of an aperture, the FT corresponding to the diffuser may be approximated by a comb function written as follows:

$$\tilde{d}(u, v) = \sum_{i=1}^{M} \sum_{j=1}^{N} \delta(u - u_i, v - v_j) \tag{9.20}$$

We have assumed a square matrix of dimensions $M \times N$ pixels, hence $M = N$.

By substituting Eq. (9.20) into Eq. (9.19), we finally obtain a repeated pattern of PSF arranged randomly over the speckle image. Hence, the complex amplitude of the speckle pattern is written as follows:

$$\tilde{B}(\rho) = \left\{\sum_{i=1}^{M} \sum_{j=1}^{N} \tilde{H}_{mn}(u - u_i, v - v_j) \otimes 2\left\{\frac{J_1(W)}{W} - \epsilon^2 \frac{J_1(W_1)}{W_1}\right\}\right\} \tag{9.21}$$

Equation (9.21) represents the repeated PSF of the HG_{mn} annular aperture and the FWHM is deduced from the speckle pattern approximately representing the speckle size. Consequently, the speckle size is nearly equal to the FWHM of the PSF. This result can be applied to any aperture.

9.4 Results and Discussion

The intensity distribution corresponding to the Hermite Gaussian laser beam for different transverse modes $HG_{n,0}$, where $n \in [1, 5]$ shown in Fig. 9.1 is computed using Eqs. (9.3–9.7). In this study, annular apertures are selected where the internal radii for the modes $= 48, 60, 72,$ and 80 pixels and the annular width $= 12$ pixels. Hence the outer radii $=$ internal radii $+$ annular width and the intensity distribution are plotted in Fig. 9.2. We compute the speckle images originating from this annular Hermite Gaussian aperture when incident on a diffuser using Eq. (9.13), and we obtain the images shown in Fig. 9.3. All the speckle images are different as expected because the PSF for the annular apertures is affected by the nonuniformity of the beam. We used the same random distribution of diffusers in the formation of all speckle images.

A comparison between the speckle images formed using $HG_{0,1}$ with an annular aperture and without an aperture is shown in Fig. 9.4. The speckle image in the LHS in the absence of the annular aperture shows elongation normal to the column number of the modes giving rise to three spots in the x-direction. Hence, the elongation occurs along the y-direction as shown in Fig. 9.4b. The other speckle image in Fig. 9.4b has spiral and random distributions as the elongation is affected by the presence of the annulus. In addition, the plots of the speckle images in both cases exhibit marked differences as shown in Fig. 9.4c. In addition, more randomness in the case of the annulus is shown in the plots which is attributed to the legs appearing in the diffraction pattern of the annular aperture.

The PSF computed for the Hermite Gaussian laser beam is obtained by operating the FFT technique upon the complex amplitude of the laser beam using Eq. (9.1). Hence, in the absence of the annular aperture, we obtain the results shown in Fig. 9.5 for different transverse modes $HG_{n,0}$, where $n \in [1, 5]$. The FWHM is deduced from the obtained PSF plots to approximately represent the speckle size and is tabulated in Table 9.1. Comparative previous results given in Ref. [18] are shown in Table 9.2. The spot size is computed from the square root of the rectangular area surrounding the HG modes. Theoretical values are calculated using Eq. (9.18) and are tabulated in Table 9.2, where $\lambda = 633$ nm, $w_0 = 300$ μm, and $f = 100$ mm.

Agreement can be seen between the speckle sizes computed from Tables 9.1 and 9.2 except that a variance occurred for $n = 1$ and 2 which is attributed to the measurement accuracy. The theoretical speckle size ranges from nearly 58–80 μm as shown in Table 9.2, while the values extracted from the FWHM range from 53 to 110 μm. Hence, the FWHM of the PSF and the speckle size is nearly equal.

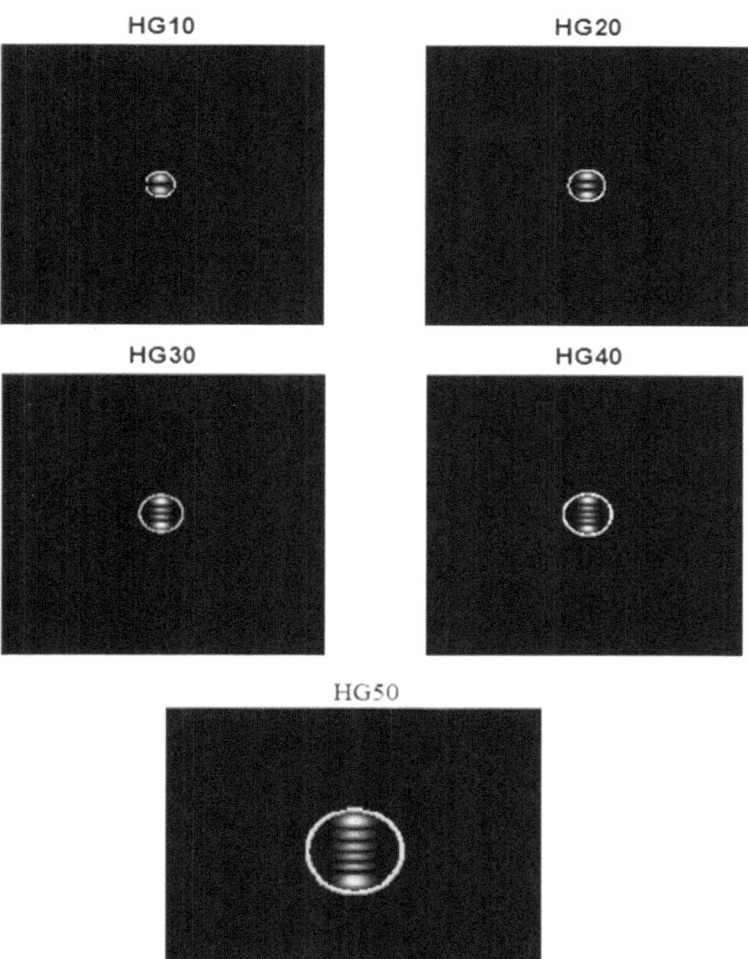

Fig. 9.2 Intensity distributions corresponding to the Hermite Gaussian laser beam for different transverse modes $HG_{n,0}$, where $n \in [1, 5]$. The beam is surrounded by an annular aperture. The internal radii are 48, 60, 72, and 80 pixels which are selected to be tangential to the outer modes

The PSF computed for the above modes of the Hermite Gaussian but surrounded by annular apertures of radii touching the outer surface of modal spots are plotted in Fig. 9.6. The computer FWHMs are tabulated in Table 9.3. The FWHM is nearly considered the average speckle size corresponding to the spots of the HG modes shown in Fig. 9.2. The corresponding spot sizes are computed by taking the square root of the internal radii shown in Fig. 9.2.

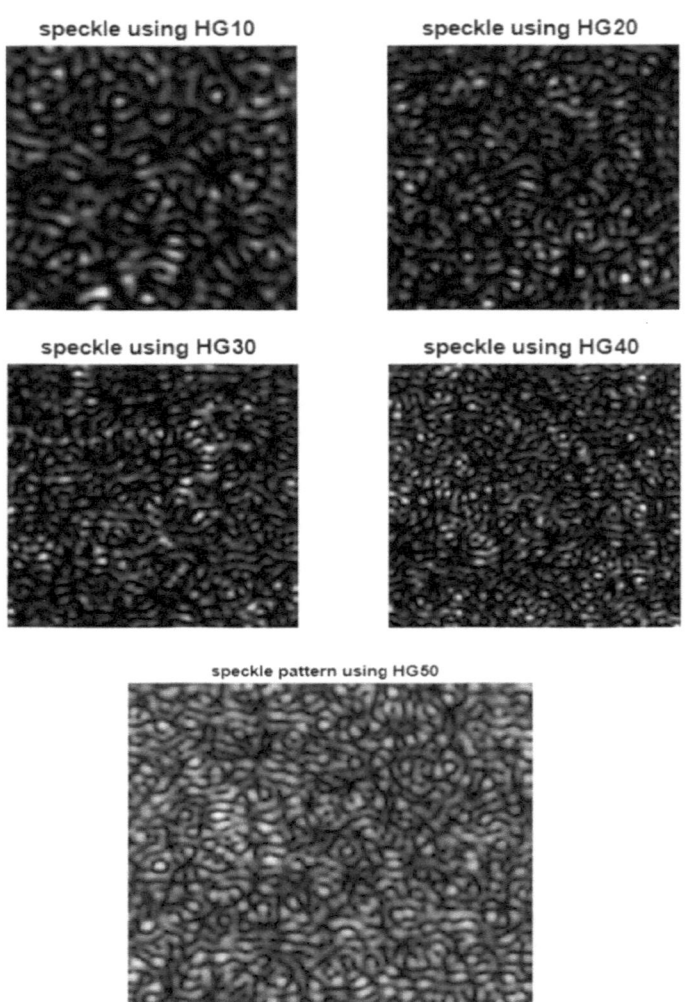

Fig. 9.3 Speckle images produced from a diffuser limited by the annular aperture and illumination with HG_{mn} the laser beam of intensity distribution is shown in Fig. 9.2

Hence, spot size $= 2w = 2\beta \sqrt{\pi (r)^2_{\text{internal}}} = 2\beta r_{\text{internal}}\sqrt{\pi}$, where 1 pixel corresponds to 10 μm. In this experiment, $r_{\text{internal}} = 0.9r_{\text{external}}$. $\beta = 0.5$ as the spot sizes surrounding the HG modes are smaller than the cross-sectional area.

Consequently, the relation between the speckle size and the spot size which is related to the numerical aperture is given for this annular HG aperture as follows:

$$\text{Speckle size}(S) = \frac{\lambda}{\text{NA}} = \frac{\lambda}{(w/f)} = \frac{\lambda f}{\beta\sqrt{\pi}(D_{\text{internal}})} \qquad (9.22)$$

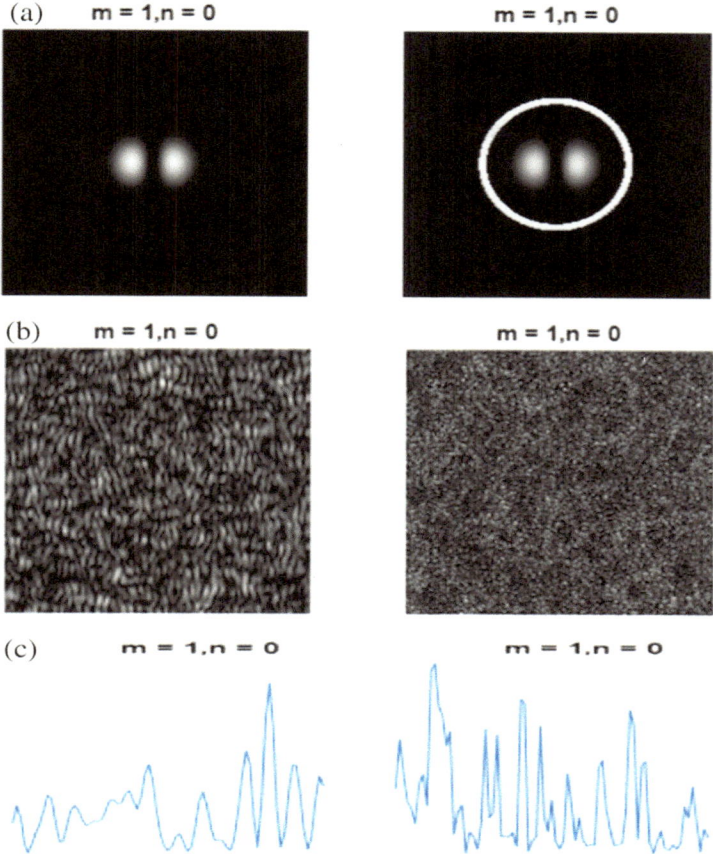

Fig. 9.4 **a** On the left, the intensity distribution for the $HG_{0,1}$ aperture is shown; on the right, the $HG_{0,1}$ beam is surrounded by an annulus with width $= 25.6$ pixels and aperture radius $= 256$ pixels, **b** Speckle pattern corresponding to the apertures shown in **a**, and **c** Plot at the horizontal line at 128 pixels in the range [16: 96 pixels]. The plots correspond to the speckle images shown in **b**

where $D_{\text{internal}} = 2r_{\text{internal}}$ is the diameter of the internal circle.

The speckle size computed from Formula (9.22) for the Hermite Gaussian annular aperture ranges from 49 to 80 μm compared with the results obtained for the absence of the annular aperture presented in Table 9.1.

A 3D plot of the speckle intensity obtained using a diffuser and illuminated with HG_{nm} annular apertures of different modes, where $m = 0$ and $n \in [1, 5]$ is shown in Fig. 9.7. A 2D plot of the speckle intensity in the absence of aperture using HG_{10} beam is shown in Fig. 9.8a, while the corresponding 3D plot is shown in Fig. 9.8b. Another plot for 2D and 3D corresponding to the speckle image produced in the case of HG_{20} illumination is shown in Fig. 9.9a, b, respectively.

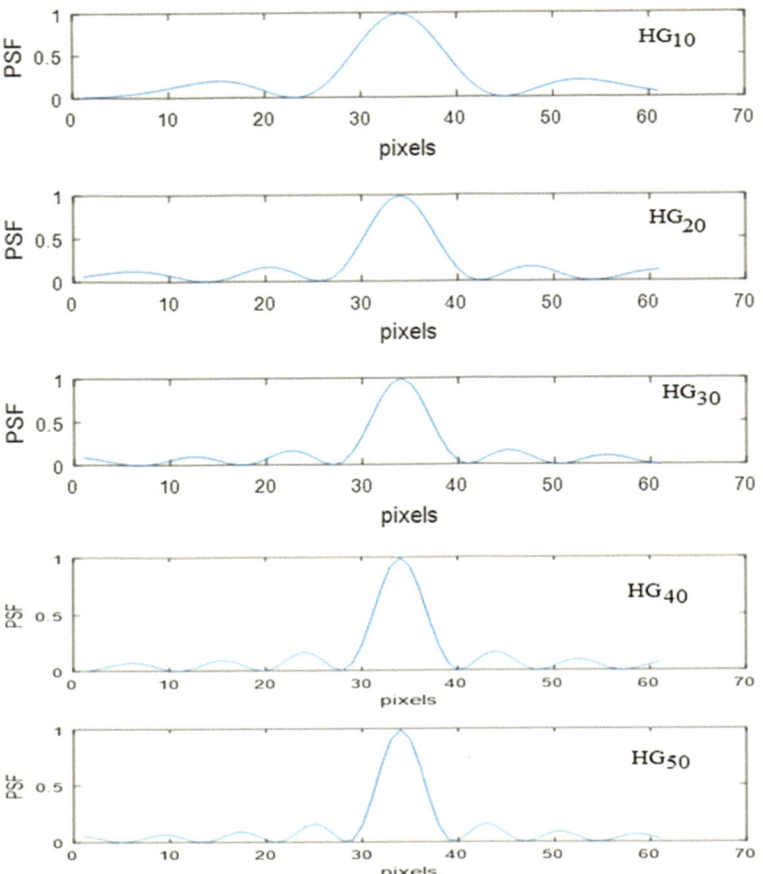

Fig. 9.5 PSF corresponding to the Hermite Gaussian laser modes. FWHM = 110, 79, 69, 60 and 53 μm, where 1 pixel = 10 μm

Table 9.1 Spot size and the corresponding speckle size using PSF for HG nm laser beam

Hermite Gaussian aperture	Spot size in mm, $w_0 = 300$ μm	Speckle size = FWHM in μm $\lambda = 633$ nm, $f = 100$ mm
$n = 1, m = 0$	0.770	110
$n = 2, \ m = 0$	0.920	79
$n = 3, \ m = 0$	0.930	69
$n = 4, \ m = 0$	1.040	60
$n = 5, \ m = 0$	1.110	53

Table 9.2 Spot size and the corresponding speckle size computed from Eq. (9.18), using HG$_{nm}$ laser beam

Hermite Gaussian aperture	Spot size in mm, $w = r_{internal}\sqrt{\pi}$	Speckle size = FWHM in μm $\lambda = 633$ nm, $f = 100$ mm
$n = 1, m = 0$	0.789	80.23
$n = 2, m = 0$	0.897	70.57
$n = 3, m = 0$	0.976	64.86
$n = 4, m = 0$	1.039	60.92
$n = 5, m = 0$	1.092	57.97

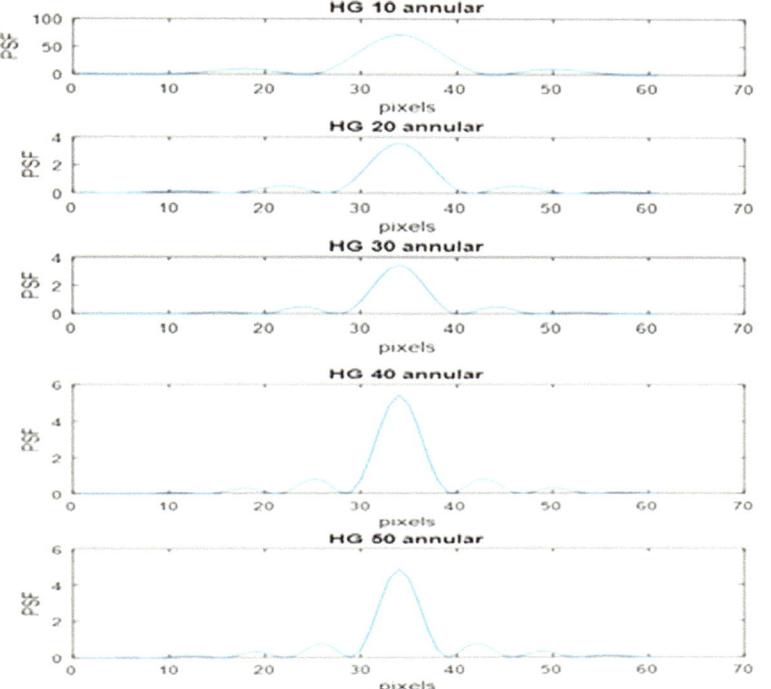

Fig. 9.6 PSF corresponding to the annular Hermite Gaussian aperture. FWHM $= 100, 87, 78, 74,$ and 69 μm, where 1 pixel $= 10$ μm

The autocorrelation along the x-direction corresponding to the speckle images using HG$_{nm}$ and annular HG$_{nm}$ apertures, where $n = 1$ and $m = 0$ are computed and plotted using MATLAB, as shown in Fig. 9.10a. The autocorrelation computed along the y-direction corresponding to the speckle images shown in Fig. 9.10a is plotted in Fig. 9.10b. The autocorrelation along the x- and y-directions corresponding to the speckle image using the HG$_{10}$ beam in the absence of the annulus is shown in

Table 9.3 Spot size and the corresponding speckle size using annular HG_{nm} aperture

HG_{nm} annular aperture	$D_{internal}$ in μm from Fig. 9.2	Spot size in mm = $\beta\sqrt{\pi}(D_{internal})$ $\alpha = 0.5$	$S = \frac{\lambda f}{\beta\sqrt{\pi}(D_{internal})}$	S = FWHM in μm $\lambda = 633$ nm, $f = 100$ mm
HG_{10}	890	0.787	80.43	100
HG_{20}	1070	0.946	66.9	87
HG_{30}	1290	1.141	55.48	78
HG_{40}	1390	1.23	51.46	74
HG_{50}	1470	1.3	48.69	69

Fig. 9.10c. FWHM = 42 pixels = 84 μm in the x-direction, while it is equal to 20 pixels = 40 μm in the y-direction.

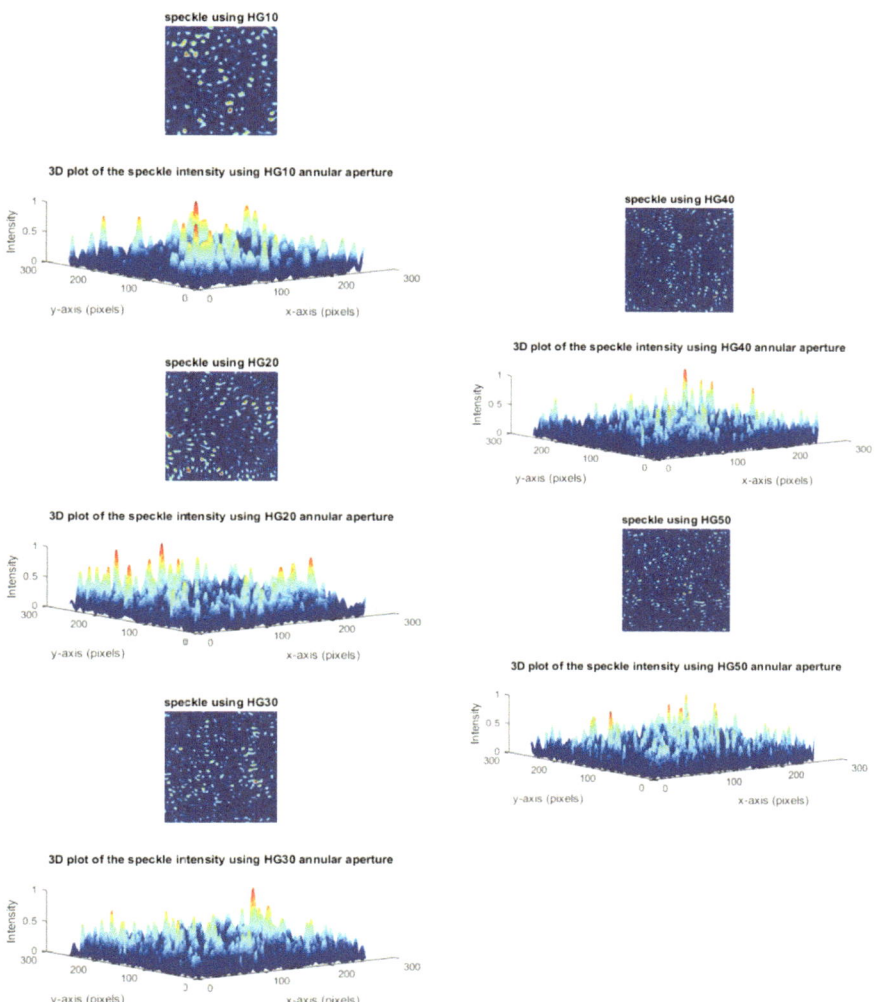

Fig. 9.7 3D plot of the speckle intensity obtained using a diffuser and illuminated with HG_{nm} annular apertures of different modes where $m = 0$ and $n \in [1, 5]$

(a)

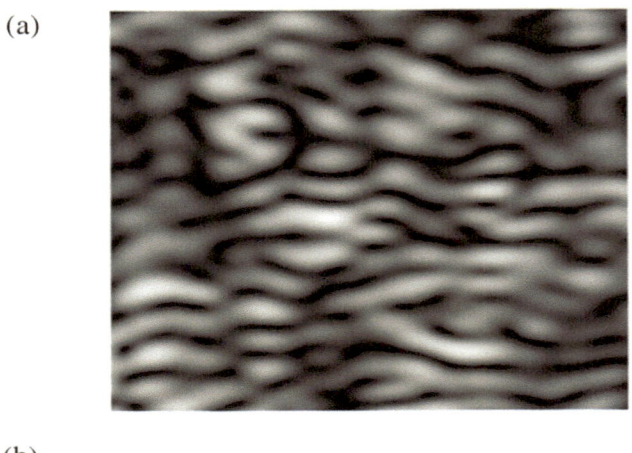

(b)

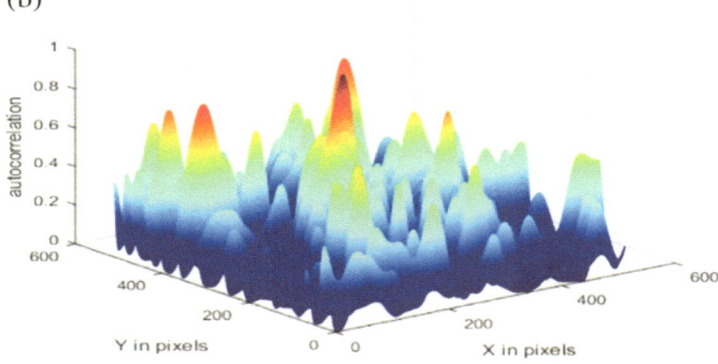

Fig. 9.8 a 2D plot of speckle intensity in the absence of an aperture using an HG_{10} beam and **b** 3D plot of speckle intensity in the absence of an aperture using an HG_{10} beam

(a)

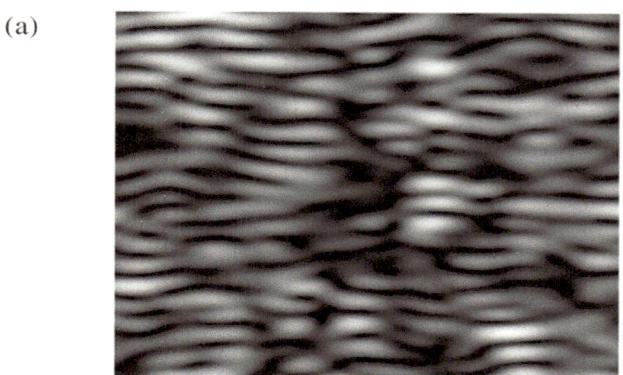

(b)

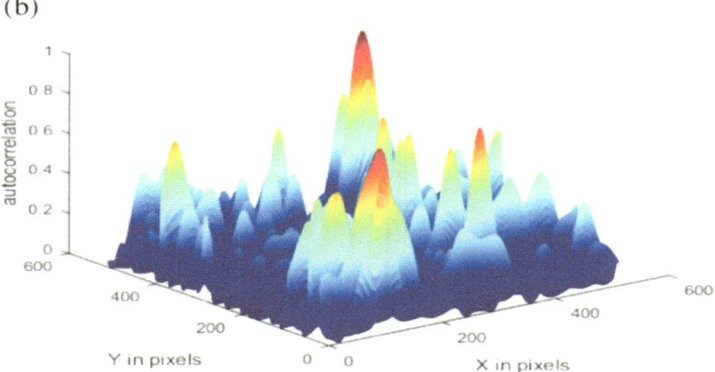

Fig. 9.9 **a** 2D plot of speckle intensity in the absence of an aperture using an HG_{20} beam and **b** 3D plot of speckle intensity in the absence of an aperture using an HG_{20} beam

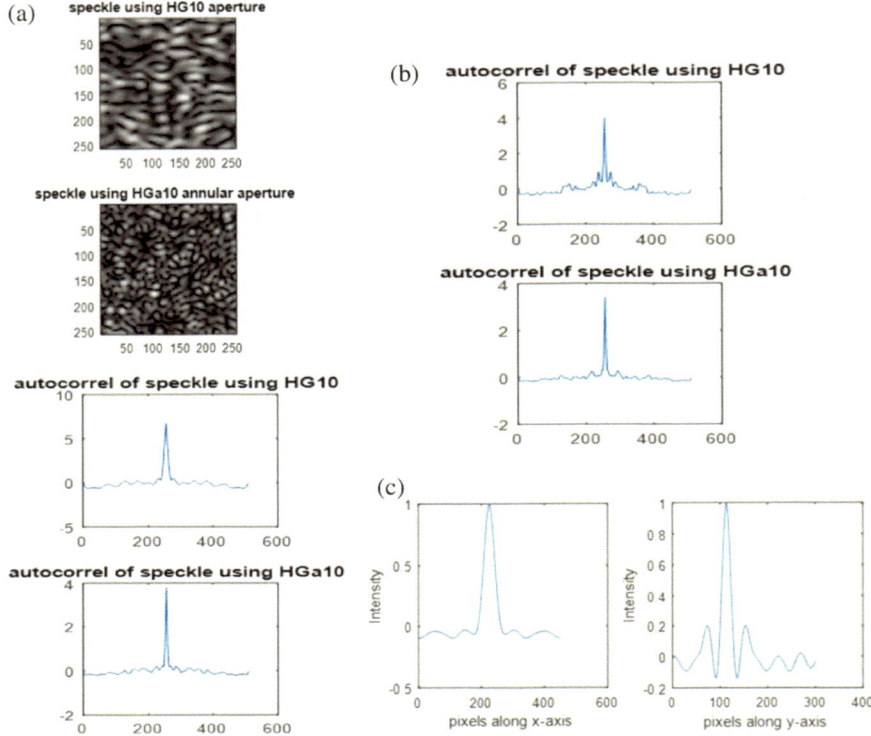

Fig. 9.10 **a** Autocorrelation along the x-direction corresponding to the speckle images using the HG_{10} beam and annular HG $_{10}$ aperture, **b** Autocorrelation along the y-direction corresponding to the speckle images shown in **a**, and **c** Autocorrelation along the x- and y-directions corresponding to the speckle image using the HG_{10} beam in the absence of the annulus. FWHM = 42 pixels = 84 μm along the x-direction, while it is equal to 20 pixels = 40 μm along the y-axis

9.5   Conclusion

First, we computed the intensity distribution for the new Hermit Gaussian annular aperture and plotted it for different transverse modes.

Second, we calculated the speckle images corresponding to these apertures using the FFT technique, which shows the dependence of the speckle pattern upon the beam nonuniformity. In addition, the elongation of the speckle structure is normal to the line containing the modes (the case of absence of the annulus) while the aperture spiral randomness of the speckle is observed by the naked eye. We confirm this reasoning because the speckle image results from the convolution of the speckle image in the case of uniform illumination with the PSF of the aperture.

Finally, the PSF is computed for both the presence and absence of an annular aperture using a Hermit Gaussian laser beam. The FWHM decreases with increasing mode number in both cases. It is deduced that the FWHM of the PSF is nearly equal to the speckle size, and a formula relating the speckle size and the spot size is deduced. In addition, the speckle size is deduced using autocorrelation of speckle images.

References

1. J.C. Dainty, in *Laser Speckle and Related Phenomena*, vol. 9 (Springer, Berlin, 1975)
2. V. Bianco, P. Memolo, et al., Strategies for reducing speckle noise in digital holography. Light. Sci. Appl. **7**, 48 (2018)
3. A.M. Hamed, Discrimination between speckle images using diffusers modulated by some deformed apertures: simulation. J. Opt. Eng. **50**, 1 (2011)
4. B. Ruffing, Digital speckle correlation for on-line real-time measurement. J. Opt. Soc. Am. A **2**, 1297 (1986)
5. P. Lehmann, Application of modulation measurement profilometry to objects with surface holes. Appl. Opt. **38**, 1144 (1999)
6. J.C. Ramirez-San-Juan, R. Ramos-Garcia et al., Impact of velocity distribution assumption on simplified laser speckle imaging equation. Opt. Express **16**, 3197 (2008)
7. J.C. Ramirez-San-Juan, E. Mendez-Aguilar et al., Effects of speckle/pixel size ratio on temporal and spatial speckle-contrast analysis of dynamic scattering systems: implications for measurements of blood-flow dynamics. Biomed. Opt. Express **4**, 1883 (2013)
8. M. Francon, *Laser Speckle and Applications in Optics* (Science Direct, 1979)
9. H. Cheng, T.Q. Duong, Simplified laser speckle-imaging analysis method and its application to retinal blood flow imaging. Opt. Lett. **32**, 2188 (2007)
10. A. Fercher, J. Briers, Flow visualization using single-exposure speckle photography. Opt. Commun. **37**, 326 (1981)
11. W. Heeman, K. Dijkstra et al., Application of laser speckle contrast imaging in laparoscopic surgery. Biomed. Opt. Express **10**, 2010–2019 (2019)
12. P. Zakharov, A. Völker et al., Dynamic laser speckle imaging of cerebral blood flow. Opt. Express **17**, 13904 (2009)
13. A.K. Dunn, A. Devor et al., Simultaneous imaging of total cerebral hemoglobin concentration, oxygenation, and blood flow during functional activation. Opt. Lett. **28**, 28–30 (2003)
14. S.G. Reddy, S. Prabhakar et al., Higher order optical vortices and formation of speckles. Opt. Lett. **39**, 4364 (2014)
15. V. Kumar, B. Piccirillo et al., Structuring Stokes correlation functions using vector-vortex beam. Opt. Lett. **42**, 466 (2017)
16. X.B. Hu, M.X. Dong, et al., Does the structure of light influence the speckle size? Sci. Rep. **10**(199), 1 (2020)
17. A.M. Hamed, Image processing of Ramses II statue using speckle photography modulated by a new Hamming linear aperture. Pramana J. Phys. **94**, 1 (2020)
18. A.M. Hamed, in *Topics on Optical and Digital Image Processing Using Holography and Speckle Techniques*. www.Lulu.com. ISBN: 9781329328464 (2015)
19. A.M. Hamed, in *The Point Spread Function of Some Modulated Apertures (Application on Speckle and Interferometry Images)* (Lambert Academic Publishing, 2017), www.lap.com, ISBN 9786202070706
20. A.M. Hamed, Formation of speckle images formed for diffusers illuminated by modulated apertures (circular obstruction). J. Mod. Opt. **56**, 1633 (2009)